Laboratory Techniques for the Detection of Hereditary Metabolic Disorders

Laboratory Techniques for the Detection of Hereditary Metabolic Disorders

Author:

VIVIAN E. SHIH, M.D.
Assistant Professor of Neurology
Massachusetts General Hospital
Boston, Massachusetts

published by:

18901 Cranwood Parkway, Cleveland, Ohio 44128

Copyright © 1973 The Chemical Rubber Co.

Second printing November 1974
CRC Press, Inc.

International Standard Book Number 0-87819-026-0

Library of Congress Catalog Card Number 73-78164

PREFACE

The application of biochemical techniques for the detection of abnormal compounds in metabolic disorders has become increasingly popular in the past decade. The purpose of this book is to provide a disease-oriented rather than technique-oriented approach, and emphasis is on the detection rather than the extensive study of the disease. Only the commonly used methods are described here. No attempt has been made to cover all the laboratory tests for special problems. There are several books dealing primarily with the techniques of chromatography of physiologic fluid which are referenced in Chapter 1.

I was fortunate to have worked with Mary L. Efron, a most inspiring teacher, who left us a wealth of knowledge, both published and unpublished, and to have been associated with the dynamic Massachusetts metabolic disorder screening program directed by my colleague, Harvey L. Levy. Experiences gained from both these laboratories are reflected in this book.

Over the past few years investigators from various countries who have visited our laboratory have often asked us to write out the routine procedures for screening for metabolic disorders. I hope that this book will be useful to those interested in this field.

I would like to thank Cynthia Valeri and Miriam Carney for their valuable assistance in the editing and preparation of the manuscript, and Petrana Peneva, Evie Barkin, and Phyllis Madigan for their excellent technical assistance in the preparation of the illustrations.

This work was supported in part by the U.S. Public Health Service Research Grant NS-05096.

THE AUTHOR

Vivian E. Shih, born on December 27, 1934, in China, graduated from National Taiwan University College of Medicine in 1958. She then came to the United States. After having completed an internship at Edgewater Hospital, Chicago, in 1959 and a residency training in pediatrics at Philadelphia General Hospital, Philadelphia, from 1960 to 1962, she took a fellowship in pediatric neurology at the Children's Hospital of Philadelphia from 1962 to 1965 and furthered her training in the field of amino acid metabolic disorders with Dr. Mary L. Efron at the Massachusetts General Hospital from 1965 to 1967. She has since engaged in the research in inborn errors of metabolism, both the clinical and biochemical aspects. She is now an assistant professor in neurology at Harvard Medical School in charge of the amino acid disorders laboratory, Joseph P. Kennedy, Jr. Memorial Laboratories, Neurology Service, Massachusetts General Hospital, and a consultant to the Massachusetts Metabolic Disorders Program, Institute of Laboratories, Department of Public Health of Massachusetts.

Dr. Shih is certified by the American Board of Pediatrics, and is a member of the Society for Pediatric Research and the Massachusetts Medical Society.

TABLE OF CONTENTS

Chapter 1

INTRODUCTION AND SUMMARY OF CLINICAL FINDINGS IN INBORN ERRORS OF METABOLISM

INTRODUCTION

The study of inborn errors of metabolism is a rapidly growing field. Hereditary metabolic disorders have been discovered by leaps and bounds in recent years; this is largely attributable to the development of simple laboratory techniques. In the mid-1940's Dent introduced paper chromatographic techniques for the study of amino acids in urine[1] to clinical medicine, a milestone in the history of biochemical genetics. Using this technique, Professor Dent's group discovered several amino acid metabolic disorders and aroused interest in these problems. Since the early 1960's there has been a great impetus toward the research of mental retardation. With improvement of the chromatographic techniques and development of new microbiological assays, large-scale screening for amino acid abnormalities became possible. This was conducted initially among the mentally retarded population and patients with neurologic abnormalities, and resulted in the discovery of many previously unknown metabolic defects.

More recent developments in the gas-liquid chromatographic techniques for organic acid analysis have contributed to the discovery of several disorders in organic acid metabolism. Wider application of such techniques will no doubt reveal additional metabolic disorders.

The application of biochemical analyses to the study of metabolic disorders has also helped to distinguish the different types of "variants" of a clinical syndrome as different diseases; the group of mucopolysaccharidoses is a good example.

Techniques for the analysis of lipids and the enzymes in their metabolism are complicated and time-consuming and are, therefore, not suitable for general screening for lipid storage diseases, except when they are performed in a specialized laboratory. Testing for metabolic abnormalities in patients with an undiagnosed disease has become an accepted medical practice. The correct diagnosis of these disorders allows the physician to employ the appropriate therapeutic measures and genetic counseling. Furthermore, the newborn screening programs for the early detection of these metabolic disorders are instrumental in preventing development of brain damage.

This book will deal primarily with the methods of screening for disorders affecting the nervous system and for which treatment is available. Only techniques that can be set up at a "screening laboratory" are described. Those which are complicated and are rarely used at a general screening laboratory will not be included.

There are several books on the basic principles and specific techniques of the various chromatographic methods frequently used in screening; these are given as general references.[2-8]

SUMMARY OF CLINICAL FINDINGS IN INBORN ERRORS OF METABOLISM

Tables 1-1 to 1-6 summarize the main clinical features and abnormal metabolites found in various metabolic disorders.[9,10] It should be cautioned that the clinical picture may be somewhat biased due to the fact that many of these disorders were discovered by screening patients with mental retardation, neurological diseases, or other problems, and may not be representative of the whole spectra of the disorder.

Table 1-7 lists the metabolic disorders that can manifest themselves as "acute" problems in the neonatal period or in the first few weeks of life. Many of the symptoms and signs are nonspecific and may be mistakenly diagnosed as birth injury or sepsis. If the underlying metabolic cause is not recognized and proper treatment not given, a fatal course may result. These metabolic disorders should always be considered in the differential diagnosis in an acutely ill patient.[11]

TABLE 1-1

Amino Acid Metabolic Disorders

Disorder	Clinical findings	Abnormal metabolites
Aromatic amino acids: Phenylketonuria (PKU) (classical or typical)	Mainly mental retardation; eczema and a musty odor may be present.	Markedly increased phenylalanine in blood ($>$ 20 mg/dl) and urine. Increased phenylpyruvic acid, phenylacetic acid, phenyllactic acid, and O-hydroxyphenylacetic acid in urine.
Atypical PKU (Severe hyperphenylalaninemia)	Asymptomatic or mental retardation.	Moderate to marked increase in blood phenylalanine (15 to 20 mg/dl). Phenolic acid derivatives may be present.
Persistent mild hyperphenylalaninemia	Asymptomatic.	Mildly increased phenylalanine in blood ($<$ 12 mg/dl) and urine.
Tyrosinosis (Hereditary tyrosinemias)		Increased tyrosine in blood, more marked in the type without hepatorenal dysfunction.
without liver disease	One case with myasthenia gravis. Other cases with mental retardation.	Increased tyrosine and its metabolites (p-hydroxyphenylacetic acid, and p-hydroxyphenyllactic acid) in urine.
with liver disease and renal tubular dysfunction	Failure to thrive, vomiting, and diarrhea, rickets, hepatosplenomegaly.	Fructosuria may be present in patient with tubular dysfunction.
Tryptophanuria with dwarfism	Developmental retardation, dwarf, photosensitivity, ataxic gait (1 case).	Borderline increased tryptophan in blood. Increased tryptophan in urine.
Hydroxykynureninuria	Bloody diarrhea, hemolytic anemia, hepatosplenomegaly, mental retardation (1 case).	Kynurenine, 3-hydroxykynurenine, xanthurenic acid.
Histidine and imidazole-dipeptide histidinemia	Asymptomatic, also found in mentally retarded patients.	Increased histidine in blood and urine; imidazole derivatives of histidine (imidazole-acetic acid and imidazole-lactic acid) in urine.
Carnosinemia and carnosinuria	Mental retardation, seizures.	Increased carnosine in urine.
Sulfur amino acids: Homocystinurias Cystathionine synthase deficiency	Dislocated lenses, mental retardation arachnoedactyly, skeletal abnormalities, thromboembolic phenomena.	Increased methionine, homocystine, and homocystine-cysteine disulfide and other derivatives of sulfur amino acids in blood and urine.
B_{12} Coenzyme metabolic defect	Failure to thrive, early death. May be asymptomatic.	Increased homocystine, homocysteine-cysteine disulfide and cystathionine in blood and urine. Normal or low blood methionine. Methylmalonic acid in urine.
$N^{5,10}$-Methylenetetrahydrofolate reductase deficiency	Seizures, muscle weakness (1 case). Episodic schizophrenic attacks (1 case).	Amino acid changes same as above except no methylmalonic aciduria.

TABLE 1-1 (Continued)

Disorder	Clinical findings	Abnormal metabolites
Cystathioninuria (Cystathionase deficiency)	Asymptomatic.	Cystathionine in blood and urine.
β-mercaptolactate-cysteine disulfiduria	Mental retardation (1 case).	β-mercaptolactate-cysteine disulfide and (?) low taurine in urine.
Sulfite oxidase deficiency	Mental retardation, decerebrated posture, dislocated lenses (1 case).	S-sulfocysteine, sulfite, and thiosulfate in urine.
Branched chain amino acids:		
Maple syrup urine disease	Early onset of seizures, feeding difficulties, ketosis, mental retardation, the presence of maple syrup odor. Several variants from mild to severe degree are known. Patients with the intermittent form have episodic occurrences of symptoms.	Increased leucine, isoleucine, alloisoleucine, and valine in blood and urine. Increased branched chain α-keto acids in urine. Above metabolites appear only during "attacks" in the intermittent form.
Hypervalinemia	Vomiting, failure to thrive, nystagmus, and hyperkinesia (1 case).	Increased valine in blood and urine.
Urea cycle and ammonia metabolism:		
Carbamyl phosphate synthetase deficiency	Early onset of vomiting, lethargy, intolerance to milk, acidosis, marked hyperammonemia (1 case).	Markedly increased blood ammonia. Increased blood and urine glutamine.
Ornithine carbamyltransferase deficiency	Early onset of episodic vomiting, lethargy, intolerance to formula, mental retardation, marked hyperammonemia. Death in neonatal period.	Markedly increased blood ammonia. Increased blood and urine glutamine.
Citrullinemia	Mental retardation. Hyperammonemia. Death in neonatal period.	Increased citrulline in blood and urine. Increased blood ammonia.
Argininosuccinic aciduria	Three types: 1. Neonatal onset of seizures, respiratory distress, progressive lethargy, and early death. 2. Enlarging liver, intolerance to formula, seizures, mental and physical retardation, trichorrhexis nodosa. 3. Mental retardation, seizures, and trichorrhexis nodosa.	Argininosuccinic acid in blood and urine. Intermittent postprandial hyperammonemia.
Hyperargininemia	Mental retardation and seizures (2 siblings reported).	Increased arginine and ammonia in blood. Increased urinary arginine, ornithine, lysine, and cystine (a pattern similar to cystinuria).
Hyperornithinemias Type I	Myoclonic seizures and mental retardation (1 case).	Markedly increased ornithine and ammonia in blood. Increased homocitrulline in urine.
Type II	Liver disease, renal tubular dysfunction and mental retardation (1 family).	Mildly increased blood ornithine.
Familial protein intolerance	Vomiting, diarrhea, growth failure, hepatosplenomegaly, hyperammonemia associated with a protein-rich diet.	Markedly increased lysine and slightly increased arginine in urine.

TABLE 1-1 (Continued)

Disorder	Clinical findings	Abnormal metabolites
Lysine:		
Hyperlysinemia	Mental retardation and synophrys.	Increased lysine and homoarginine in blood. "Cystinuria" pattern, pipecolic acid, homoarginine and acetyl-lysines in urine.
Congenital lysine intolerance	Early onset of vomiting, convulsion, coma, and intolerance to protein feeding (1 case).	Periodic increase in blood lysine and arginine. Hyperammonemia.
Saccharopinuria	Mental retardation, dwarfism. Spastic diplegia.	Increased saccharopine, homocitrulline, lysine, and citrulline in blood. Increased saccharopine, lysine, homocitrulline, citrulline, homoarginine, and aminoadipic acid in urine.
Hyperpipecolatemia	Feeding problem, hepatomegaly, progressive mental retardation, hypotonia and nystagmus (1 case).	Pipecolate in blood, very little in urine.
Hydroxylysinuria	Mental retardation and seizures.	Hydroxylysine and acetyl-lysines in urine.
Glycine:		
Nonketotic hyperglycinemia	Mental retardation, hyperactivity.	Increased glycine in blood and urine.
Ketotic hyperglycinemia (Propionic acidemia)	Ketosis of early onset, growth and development retardation.	Increased glycine and propionic acid in blood and urine.
Sarcosinemia	Mental retardation.	Sarcosine in blood and urine.
Imino acids:		
Hyperprolinemia		
Type I	Mental retardation, seizures, renal disease.	Increased proline in blood. Increased proline, hydroxyproline, and glycine in urine.
Type II	Mental retardation, seizures.	Increased proline in blood. Increased proline, hydroxyproline, and glycine in urine. In addition, increased Δ-pyrroline-5-carboxylate in urine.
Hydroxyprolinemia	Probably asymptomatic.	Hydroxyproline in blood and urine.
Miscellaneous:		
β-alaninemia	Uncontrollable seizures. Somnolence (1 case).	Increased β-alanine and γ-aminobutyric acid in blood; increased β-alanine, β-aminoisobutyric acid, γ-aminobutyric acid and taurine in urine.
Hypophosphatasia	Rickets or osteomalacia.	Phosphoethanolamine.
Pseudohypophosphatasia	Rickets or osteomalacia.	Phosphoethanolamine.
Hyperalaninemia pyruvate decarboxylase deficiency	Intermittent ataxia, choreoathetosis, mental retardation.	Increased pyruvate, and alanine in blood and urine during attacks.

TABLE 1-1 (Continued)

Disorder	Clinical findings	Abnormal metabolites
Pyruvate carboxylase deficiency	Progressive and relapsing episodes of anorexia, hypotonia, hyporeflexia, ataxia; mental retardation (Leigh's encephalomyelopathy).	Increased lactate and alanine in blood during attacks.
Aspartylglycosaminuria	Mental retardation, coarse features, seizures	Aspartylglycosamine in blood and urine.
Pyroglutamic aciduria	Mental retardation, episodic vomiting, spastic quadriparesis, ataxia (1 case).	Large amount of pyroglutamic acid (not detectable by amino acid screening).

TABLE 1-2

Summary of Amino Acid Transport Disorder

Disorder	Clinical findings	Abnormal metabolites
Hartnup disease (Neutral aminoaciduria)	May be entirely normal. May have intermittent ataxia, photosensitive rashes, mental retardation, psychosis.	Marked increases in urinary neutral amino acids (alanine, threonine, serine, glutamine, histidine, isoleucine, leucine, valine, phenylalanine, tyrosine, tryptophan). Methionine excretion variable. Urinary excretion of indole derivatives and stool amino acids were increased in some cases.
Methionine malabsorption	Mental retardation, convulsions with abnormal electroencephalogram. White hair, an unpleasant odor (2 cases).	Intermittent methioninuria; urinary excretion of a-hydroxybutyric acid; increased methionine in stool.
Blue diaper syndrome (tryptophan malabsorption)	Failure to thrive, recurrent infections, hypercalcemia, nephrocalcinosis, blue diaper (1 family).	Increased excretion of indole derivatives including indole-acetamide, indolactic acid, indole acetylglutamine, and indican.
Cystinuria (Type I to III)	Urinary stone.	Increased cystine, lysine, ornithine, and arginine. Heterozygotes of types II and III have increased cystine lysinuria.
Isolated cystinuria	Hypoparathyroidism hypocalcemic tetany.	Increased urinary cystine only.
Dibasic aminoaciduria	Asymptomatic.	Lysine, arginine, and ornithine.
Iminoglycinuria	Asymptomatic.	Increased urinary excretion of proline, hydroxyproline, and glycine; heterozygotes have increased urinary glycine.
Miscellaneous syndromes Fanconi syndrome	Variable, renal tubular acidosis and hypophosphatemic rickets.	Generalized hyperaminoaciduria, glycosuria.
Lowe syndrome (Oculocerebrorenal syndrome)	Mental retardation, cataract and other eye anomalies, renal tubular acidosis.	Generalized hyperaminoaciduria, glycosuria.
Glucoglycinuria	Asymptomatic.	Increased glucose and glycine excretion in urine.

TABLE 1-3

Disorders in Sugar Metabolism

Disorder	Clinical findings	Abnormal metabolites
Galactose:		
Galactosemia (Gal-1-phosphate transferase deficiency)	Failure to thrive, vomiting, intolerance to milk, hepatomegaly, jaundice, cataract, and mental retardation.	Galactose and galactitol in blood and urine. Galactose-1-phosphate in erythrocytes.
Galactokinase deficiency	Cataract.	Galactose and galactitol in blood and urine.
Fructose:		
Hereditary fructose intolerance (fructose-1-phosphate aldolase deficiency)	Two clinical types: (1) Early onset; vomiting, failure to thrive, hypoglycemia, bleeding tendency, renal tubular dysfunction (Fanconi's syndrome), and early death. No rickets. (2) Late onset; less severe, aversion to sweet food, hypoglycemia, hypoprothrombinemia; may be asymptomatic.	Fructosuria; generalized amino-aciduria; in some cases, hyper-methioninemia, hypertyrosinemia, and hypertyrosyluria.
Fructose diphosphatase deficiency	Feeding difficulties, failure to thrive, jaundice, hepatomegaly, hypoglycemia, and hypotonia.	Lactic acidemia.
Essential fructosuria (fructokinase deficiency)	Asymptomatic.	Fructosuria after ingestion of fructose-containing food.
Galactose and fructose:		
Familial galactose and fructose intolerance	Severe hypoglycemia induced by fructose and galactose (1 family).	Fructose and galactose, generalized increases in amino acid excretion after fructose and galactose loading.
Pentose:		
Essential pentosuria	Asymptomatic, mainly in Jewish persons.	Large amounts of xylulose, little xylose in urine.
Lactose:		
Lactose intolerance (without lactase deficiency)	Diarrhea, vomiting, cachexia, renal acidosis.	Lactose in urine and generalized aminoaciduria.
Lactose intolerance (lactase deficiency)	Failure to gain weight in infancy; dislike for milk in adults.	Lactose in urine, variable amount.
Sucrose:		
Sucrose intolerance (sucrase deficiency)	Diarrhea, intolerance to table sugar and sweets.	Sucrose in urine and stool.

TABLE 1-4

Disorders in Sugar Transport

Disorder	Clinical findings	Abnormal metabolites
Renal glycosuria	Asymptomatic.	Glucose in urine.
Glucose galactose malabsorption	Severe diarrhea, dehydration, intolerance to all dietary carbohydrates, failure to thrive.	Glucose in stool and mild glycosuria intermittently.
Glucoglycinuria	Asymptomatic.	Increased glucose and glycine in urine.

TABLE 1-5

Disorders in Organic Acids

Disorder	Clinical findings	Biochemical abnormalities
Isovaleric acidemia (isovaleryl CoA dehydrogenase deficiency)	Recurrent vomiting, acidosis, mental retardation, "sweaty feet" odor.	Isovaleric acid and isovalerylglycine in blood and urine; may have hyperglycinemia.
β-Methylcrotonylglycinuria and β-hydroxyisovaleric acidemia (β-methylcrotonyl CoA carboxylase deficiency)	Muscular hypotonia, feeding difficulties, motor retardation, peculiar odor in urine (like that of cat's urine) (1 case).	β-hydroxycrotonyl-glycine and β-hydroxyisovaleric acid in urine.
Propionic acidemia (propionyl CoA carboxylase deficiency)	Recurrent ketoacidosis, protein intolerance, mental retardation, leucopenia, thrombocytopenia, and hypogammaglobulinemia.	Increased propionic acid and glycine in blood and urine; hyperammonemia may be present.
a-Methyl-acetoacetic acidemia (β-keto-thiolase deficiency)	Recurrent ketosis, neutropenia, thrombopenia.	Increased a-methyl-acetoacetic acid, a-methyl-B-hydroxybutyric acid in blood and urine; hyperammonemia and hyperglycinemia may be present.
Methylmalonic acidemias: Defect in methylmalonyl CoA mutase	Recurrent ketoacidosis, mental retardation, recurrent infection, osteoporosis, neutropenia.	Increased methylmalonic acid and glycine in blood and urine.
Defect in B_{12}-coenzyme metabolism.	Recurrent ketoacidosis, mental retardation, recurrent infection, osteoporosis, neutropenia. Responsive to B_{12} therapy.	Increased methylmalonic acid and glycine in blood and urine.
Defect in methylmalonyl CoA racemase	Metabolic acidosis, coma, and neonatal death (1 case).	Increased methylmalonic acid in blood and urine; hyperammonemia.
Pyruvic acidemia and hyperalaninemia (pyruvate decarboxylase deficiency)	Intermittent ataxia, choreoathetosis, mental retardation.	Increased pyruvate, alanine in blood and urine, more marked during attacks; slight increase in lactic acid; presence of inhibitor of thiamine triphosphate synthesis in blood and urine.
Lactic acidemia and hyperalaninemia (pyruvate carboxylase deficiency)	Subacute necrotizing encephalomyelopathy (Leigh's), progressive and relapsing muscle weakness, hypotonia, hyporeflexia, ataxia, seizures; may be retarded.	Increased lactic acid and alanine in blood and urine; presence of inhibitor of thiamine triphosphate synthesis.
Pyroglutamic aciduria	Mental retardation, episodic vomiting, spastic quadriparesis, ataxia (1 case).	Large amount of pyroglutamic acid.

TABLE 1-6

The Mucopolysaccharidoses (Classification of McKusick)

Disorder		Clinical findings					Abnormal metabolites in urine
Type	Syndrome	Mental retardation	Skeletal deformities	Corneal opacity	Genetics	Hepatosplenomegaly	
I	Hurler's	Severe	Marked	+	Autosomal recessive	+	Dermatan sulfate Heparin sulfate
II	Hunter's	Moderate	Marked	−	Sex-linked recessive	+	Dermatan sulfate Heparin sulfate
III	Sanfilippo's	Severe	Mild	?	Autosomal recessive	+	Heparin sulfate
IV	Morquio's	−	Severe, marked spondyl-epiphyseal dysplasia	±	Autosomal recessive	+	Keratan sulfate
V	Scheie's	−	Mild	Severe	Autosomal recessive	Hepatomegaly	Dermatan sulfate
VI	Maroteaux-Lamy	−	Marked	+	Autosomal recessive	+	Dermatan sulfate
VII		+	Similar to Morquio's	+	Autosomal recessive	+	Keratan sulfate-like material and dermatan sulfate

TABLE 1-7

Metabolic Disorders with Onset of Clinical Manifestations in Neonatal Period or Early Infancy

Disorders	Vomiting, acidosis	Poor feeding FTT	MR	Other neurological abnormality	Liver dysfunction	Renal tubular dysfunction	Odor
Maple syrup urine disease	+	+	+	+	−	−	Maple syrup
Ketotic hyperglycinemia	+	+	+	±	−	−	−
Nonketotic hyperglycinemia	−	−	+	+	−	−	−
Argininosuccinic-aciduria	±	+	+	+	+	−	−
Congenital hyperammonemias	+	+	+	+	+	−	−
Citrullinemia	+	+	+	+	−	−	−
Hyperornithinemia type II	−	+	+	−	+	+	−
Hereditary tyrosinemia	−	+	+	−	+	+	±
Isovaleric acidemia	+	+	+	+	−	−	Sweaty feet
Methylmalonic acidemia	+	+	+	−	−	−	−
Pyruvic and lactic acidemia	+	+	+	+	−	−	−
Galactosemia	+	+	+	+	+	+	−
Fructose intolerance	−	+	+	+	+	+	−

Abbreviations:
FTT = Failure to thrive
MR = Mental retardation

REFERENCES

1. **Dent, C. E.,** Detection of amino acids in urine and other fluids, *Lancet,* 2, 637, 1946.
2. **Smith, I.,** Ed., *Chromatographic and Electrophoresis Techniques, Volume I, Chromatography,* 3rd ed., John Wiley & Sons, New York, 1969.
3. **Smith, I.,** Ed., *Chromatographic and Electrophoresis Techniques, Volume II, Zone Electrophoresis,* 3rd ed., John Wiley & Sons, New York, 1968.
4. **Block, R. J., Durrum, E. L., and Zweig, G.,** *Paper Chromatography and Paper Electrophoresis,* 2nd ed., Academic Press, New York, 1958.
5. **Lederer, E. and Lederer, M.,** *Chromatography. A Review of Principles and Applications,* 2nd ed., Elsevier, New York, 1957.
6. **Stahl, E.,** Ed., *Thin-Layer Chromatography. A Laboratory Handbook,* Springer-Verlag, Berlin, 1965.
7. **Randerath, K.,** *Thin-Layer Chromatography,* 2nd ed., Academic Press, New York, 1966.
8. **Burchfield, H. P. and Storrs, E. E.,** *Biochemical Applications of Gas Chromatography,* Academic Press, New York, 1962.
9. **Stanbury, J. B., Wyngaarden, J. B., and Fredrickson, D. S.,** Eds., *The Metabolic Basis of Inherited Disease,* 3rd ed., McGraw-Hill, New York, 1972.
10. **Tanaka, K.,** Pathogenesis of disorders of organic acid metabolism associated with brain dysfunction, in *Biology of Brain Dysfunction,* Gaull, G. E., Ed., Plenum Press, in press.
11. **O'Brien, D. and Goodman, S. I.,** The critically ill child: Acute metabolic disease in infancy and early childhood, *Pediatrics,* 46, 620, 1970.

Chapter 2

AMINO ACIDS

INTRODUCTION

Techniques for detection of amino acids in physiological fluids are simple and dependable. The availability of these techniques has made possible programs for screening many different populations for amino acid abnormalities. As a result of these studies, over 30 inborn errors of metabolism in which amino acid abnormalities are either primary or secondary have been discovered. In a number of these disorders, abnormal metabolites other than amino acids are also excreted. The finding of these compounds by preliminary simple chemical tests would be a clue to the existence of amino acid abnormalities and leads to further studies. For convenience, these chemical tests are included in this chapter and described below.

CHEMICAL TESTS

Many of the chemical tests for the detection of abnormal metabolites of amino acids are nonspecific and react with a group of substances as a screening test should. Several commonly used screening tests are described below.

Ferric Chloride Test
   Reagent: 10% ferric chloride in water.
   Method: To 1 or 2 ml of urine add ferric chloride solution drop by drop.
   Results and interpretation: Many compounds react with ferric chloride to form various color complexes (Table 2-1). In making a presumptive diagnosis of PKU, a green ferric chloride test should be supported by a positive dinitrophenylhydrazine test. Although a color reaction to ferric chloride reagent has been observed with histidinemia and tyrosinemia, results in our laboratory have been inconsistent and unreliable. A negative ferric chloride test does not necessarily rule out any of these disorders.
   A commercially prepared filter paper strip impregnated with buffered ferric chloride* will

TABLE 2-1

Color Reaction of Various Compounds with Ferric Chloride Reagent

Disorders or compounds	Color reaction with ferric chloride	Color reaction with Phenistix
Phenylketonuria (Phenylpyruvic acid)	Green	Green
Tyrosinemia (p-hydroxyphenyl-pyruvic acid)	Rapidly fading green	Rapidly fading green
Maple syrup urine disease (Branched chain ketoacidosis)	Greenish gray	Negative
Histidinemia (Imidazolepyruvic acid)	Gray	Gray
Alkaptonuria (Homogentisic acid)	Effervescent blue	Negative
Acetoacetic acid	Reddish brown	Negative
Salicylates	Purple	Purple
p-Aminosalicylic acid	Reddish brown	Purple
Phenothiazines	Grayish green	Grayish green
Antipyrin	Deep orange	
Isoniazid (Isonicotinic hydrazide)	Gray	

*Phenistix®, Ames Company

give essentially the same information. A variety of drugs will react with $FeCl_3$ and give a positive reaction. Therefore, a careful inquiry about drug ingestion is essential when a positive result is obtained. The ferric chloride test can also be used in the diagnosis of drug poisoning.

2,4-Dinitrophenylhydrazine Test (DNPH Test)

Reagent: 0.1% 2,4-dinitrophenylhydrazine (DNPH) in 2 N hydrochloric acid.

Method: To 2 ml of the urine specimen in a test tube, add an equal amount of the DNPH reagent and mix. The mixture should be examined immediately after its addition and again 10 min later.

Results and interpretation: The appearance of a yellow precipitate 10 min after the addition of DNPH reagent indicates the presence of a-keto acids. The test is considered negative when the mixture remains clear after standing for 10 min. For confirmation, the yellow precipitate can be extracted with 5 ml of ether. The ether layer is then removed and extracted with 5 ml of 10% sodium carbonate. When 5 ml of 10% sodium hydroxide solution are added, a reddish brown color develops in the water phase. This test is a complement to the ferric chloride test in the diagnosis of PKU. Patients with a blood phenylalanine level over 16.5 mg/100 ml or 1 mmol usually have a positive test. It is much more useful in the diagnosis of MSUD. A positive reaction usually indicates that the blood leucine level is above 10 mg/100 ml or 0.8 mmol/l.

Nitrosonaphthol Test[1]

Reagents:

1. Nitric acid, 2.63 N (add 1 part of concentrated nitric acid to 5 parts water).

2. Sodium nitrite, 2.5% in water (w/v). Store at $4°C$.

3. 1-nitrosonaphthol, 0.1% in 95% ethanol (w/v). Store at $4°C$.

Method: To 1 ml of the 2.63 N nitric acid in a test tube add 1 drop (from a Pasteur pipette) of the sodium nitrite solution and then 10 drops of the nitrosonaphthol reagent. Agitate to mix (mechanical mixer may be used). Add 3 drops of urine and mix again. Read color change within 2 to 5 min.

Results and interpretation: The appearance of an orange-red color indicates the presence of excessive amounts of the derivatives of p-

hydroxyphenol such as tyrosine, p-hydroxypyruvic acid, p-hydroxyphenylacetic acid, and p-hydroxyphenyllactic acid. The Millon reaction, as described by Medes,[2] can be used for quantitative measurement of tyrosine derivatives in urine. O-Hydroxyphenolic acid derivatives, such as the metabolites in PKU, do not react with this reagent.

Cyanide-nitroprusside Test

Reagents for liquid urine specimen:

1. Sodium cyanide 5% in water (w/v). Stable for at least one month.

2. Sodium nitroprusside (sodium nitroferricyanide, Na_2 Fe $(CN)_5$ NO$\cdot$2H$_2$O) 5% in water (w/v). Prepare fresh daily as needed.

Reagents for filter paper urine specimen:

1. Dissolve sodium cyanide, 5 g in small amount of water. Add 95% ethanol to 100 ml.

2. Dissolve sodium nitroprusside crystals, 0.2 to 0.5 g in a small amount of water. Add 95% ethanol to approximately 10 ml. Make fresh as needed.

Method: To 2 ml of urine add 1 ml of 5% aqueous sodium cyanide solution. Mix and let stand for at least 5 but not more than 20 min (sensitivity decreases during longer periods). Then add nitroprusside solution drop-wise and mix.

When a filter paper urine specimen is the only one available, this test can be modified by using the reagents made up in ethanol. Punch disc from the filter paper with a 3/8 in. paper punch. Place the disc in a white porcelain hanging-drop dish. Wet the disc with a drop of 5% alcoholic sodium cyanide. Wait 5 to 10 min, and then touch the disc with the dilute alcoholic nitroprusside solution from a Pasteur pipette. It is important that only enough reagent to wet the disc be applied. An excessive amount will dilute the color intensity.

Results and interpretation: An immediate reddish-purple color indicates a positive reaction. Any compound with a disulfide linkage will give a positive reaction. Often normal concentrated urines will show a weakly positive (1+) reaction.

This test is particularly useful in the diagnosis of homocystinuria. Homocystine runs together with serine in the electrophoresis and butanol-acetic acid-water solvent systems frequently used in screening. Without the cyanide-nitroprusside test, homocystine might not be detected. A positive reaction suggests, in addition to homocystinuria, at least two other metabolic disorders in which a disulfide is excreted. These are cystinuria (Table 1-2) and β-mercaptolactate-cysteine disulfiduria (Table 1-1). A drug, *N*-acetylcysteine,* when given as an inhalant to a patient, is excreted as acetylcystine and acetylcysteine-cysteine mixed disulfide. Both compounds react to the cyanide-nitroprusside reagent.

Silver-nitroprusside Test for Homocystine[3]

Reagents: Silver nitrate 1.0% (w/v) in 3% ammonia (store in dark bottle).

Sodium nitroprusside 1% (w/v) (stable for several days at 4°C).

Sodium cyanide 0.7% (w/v).

Method and results:

1. For liquid specimen:

Saturate urine with solid sodium chloride. Transfer 5 ml of the salt-saturated urine to a test tube (A) and add to it 0.5 ml silver nitrate solution. To a control tube (B) containing 5 ml urine, add only 0.5 ml of 3% ammonia without silver nitrate. After 1 min, add 0.5 ml of nitroprusside solution to each tube; follow by 0.5 ml cyanide solution to each tube.

Immediate development of a purple color in tube (A) indicates the presence of homocystine. Cystine will not react. In the control test tube (B) cystine gives a purple color, while homocystine does not react.

2. Modified method for filter paper urine specimen:[4]

Place disc of filter paper specimen in a white porcelain hanging dish. Add 1 drop of the silver nitrate solution, just enough to wet the disc. A brownish color will develop due to the precipitation of inorganic salt by silver nitrate. After 2 min, add 1 drop of the nitroprusside solution. Let it stand for 1 min and transfer the disc to another well. Add 1 drop of the cyanide solution to the disc. Read the color between 1 and 2 min. The development of a purple color indicates the presence of homocystine. This test is sensitive to homocystine at 10 mg/100 ml. Since this test is specific for homocystine, it should not be used to replace the routine cyanide-nitroprusside test for screening. It should be used to differentiate homocystine from cystine after a positive cyanide-nitroprusside test has been obtained.

Thiosulfate Test[5]

Reagent: Sodium azide 3% in 0.1 *N* iodine (w/v).

Method: Place either 1 drop of urine or a disc of filter paper urine specimen in a white porcelain hanging dish. Add 1 drop of the reagent.

Results and interpretation: Immediate bubbling and decolorization indicates a positive reaction. A patient with sulfite oxidase deficiency excretes an excessive amount of thiosulfate, as well as sulfite and S-sulfocysteine. Since sulfite is unstable, the presence of thiosulfate may prove to be more reliable in making the diagnosis.

Blood Ammonia Determination

Blood ammonia determination is an often neglected but very useful screening test for amino acid metabolic disorders. Increased glutamine is the only amino acid abnormality in carbamyl phosphate synthetase deficiency and ornithine transcarbamylase deficiency. The diagnosis of these two urea cycle disorders, therefore, depends upon the finding of marked hyperammonemia. There are also several other congenital disorders (Tables 1-1 and 1-2) associated with hyperammonemia.

A number of procedures are available for ammonia determination. These are modifications of methods based on either microdiffusion,[6,7] ion exchange,[8,9] or an enzymatic method.[10,11] Only the modified Conway method[6] is described here. Several references to other methods are given and the authors' original papers should be consulted for details.

Reagents:

Stock solutions:

1. Indicators, methyl red 100 mg in 100 ml of 95% ethanol, and bromcresol green 100 mg in 100 ml of 10% ethanol.

2. Boric acid, 10g in 700 ml of water and 200 ml of ethanol.

3. Potassium carbonate, saturated solution.

4. Hydrochloric acid, 1 *N*.

Working solutions: These reagents should not be made until the time of use.

1. Indicator solution:

Bromocresol green (0.1%)	1.0 ml.
Boric acid (1.1%)	5.0 ml.
Methyl red (0.1%)	0.4 ml.

*Mucomyst®, Mead Johnson and Company

Dilute to 50 ml with freshly deionized ammonia-free water.

2. Hydrochloric acid, 0.0025 N. Dilute 0.25 ml of N hydrochloric acid to 100 ml with freshly deionized ammonia-free water.

Standards: Solutions of ammonium hydroxide are made up of 25, 50, 100, and 150 μg of ammonia nitrogen per 100 ml.

Method: The porcelain dishes are soaked in cleaning solution overnight, rinsed thoroughly with distilled water, and stored in a jar of very dilute sulfuric acid solution (3 ml of concentrated sulfuric acid in 4 l. of water). Before use, rinse the dishes with freshly deionized water and allow to dry upside down.

Blood should be drawn in a heparinized syringe and kept in the syringe on ice; alternatively, it can be drawn into an EDTA vacuum tube and kept on ice. The measurement should be done within 1 hr, as any further delay will cause an artifactual rise in ammonia content.

Pipette 1 ml of the freshly made indicator mixture into the center well of a porcelain Conway dish and 1 ml of saturated potassium carbonate solution into one side of the outer well. Cover the dish with a lid that has been pregreased with vaseline or silicone grease. Slide the lid to one side so that 1 ml of blood or the standard solution can be pipetted to the side opposite the potassium carbonate. Slide the lid back to make the dish airtight and slowly swirl the dish to mix the blood and potassium carbonate. Shake the dish on a rotary shaker for 30 min at a constant speed so that the blood and potassium carbonate are thoroughly mixed without spilling.

Draw 0.0025 N hydrochloric acid into a microburette syringe calibrated to deliver 0.20 μl/division. The indicator solution in the center well is titrated back to the original reddish-pink color by using the microburette.* Calculation:

$$\frac{\text{number of divisions} \times 0.0025 \times 0.02 \,\mu\text{l}}{\text{ml of blood used}} =$$

micromoles of ammonia N/100 ml blood

Micromoles of ammonia $N \times 14 = \mu$g of ammonia N

Normal range: Up to 75 μg of ammonia N/100 ml

Cautions: It is well known that results of ammonia determinations are not always stable. It is important to keep the laboratory free of ammonia contamination. Reagents containing ammonia should not be stored or used in any room where ammonia is measured. Samples should always be run in duplicate and with standards.

Hyperammonemia can be intermittent in some of the metabolic disorders. The blood ammonia level is normal in the fasting state, but becomes elevated after protein ingestion. Thus, a 2-hr postprandial ammonia determination is more meaningful than a fasting level and is recommended to rule out hyperammonemia.

Test for Reducing Substance

The Benedict test or Clinitest is primarily for detection of reducing substances and not of amino acids. Details of the method are discussed in Chapter 3.

Melituria may be either the cause of generalized aminoaciduria or a part of a genetic syndrome such as galactosemia, Fanconi syndrome, etc.

Obermayer Test[1 2]

Reagents: Ferric chloride, 0.2% in concentrated hydrochloric acid (w/v); chloroform.

Method: To 4 ml of urine add 5 ml ferric chloride reagent and mix by inverting the tube ten times, then add 2 to 3 ml of chloroform. Mix by inverting several times again.

Results and interpretation: Development of a blue color in chloroform layer indicates a positive reaction and the presence of indican in the urine. Urinary indican is of exogenous origin; it is a derivative of an absorbed bacterial breakdown product of tryptophan in the intestine. Indole produced by the intestinal bacteria from tryptophan is absorbed, oxidized to indoxyl, and conjugated with sulfate to form indican. A person who has Hartnup disease, intestinal malabsorption, or constipation may excrete increased amounts of indican in the urine. However, a negative result does not rule out Hartnup disease.

HIGH VOLTAGE ELECTROPHORESIS

High voltage electrophoresis (HVE) is suitable for the separation of small molecules such as amino acids in physiologic fluids. It has been used alone for one-dimensional separation or in combination with a solvent chromatography to give

*Micrometric Instrument Company, Cleveland, Ohio

two-dimensional separation as a screening procedure.

Electrophoresis has the advantage of eliminating the step of desalting the specimen and the procedure takes less time.

Equipment

High voltage paper electrophoresis generates large amounts of heat, and an efficient cooling system is essential for use with such apparatus.* In one type, the paper is immersed in a large volume of a water-immiscible liquid coolant. In another type, the paper is sandwiched between two layers of polyethylene sheets (for insulation) and placed on a cooling metal plate with tap water circulating in the tubes. The latter type has certain advantages over the former. It is easier to work with and the buffer can be changed very easily from one system to another. All the electrophoresis runs discussed later have been performed on this type of apparatus.

A separate unit of power supply provides variable voltages from 0 to 3,000 or 5,000 V, and the amperage up to 300 mA, depending upon the model. The unit providing 3,000 V is quite adequate for ordinary laboratory use.

Safety Precautions

A fatal accident related to the operation of a high voltage apparatus has been reported.[13] It is of utmost importance that every effort be made to install safety devices in the proper manner and to have them tested periodically by a qualified person. The units should be well grounded. Additional information about the apparatus is usually supplied by the manufacturer. Laboratory personnel should be completely familiar with the apparatus before using it.

Buffer vessels should be emptied and washed immediately after use to avoid drying up of the buffer and salt condensation. The polyethylene sheets should also be wiped clean immediately. Any salt condensation on the sheet will draw a large amount of current and generate excessive heat which may burn the polyethylene sheet and cause an electric short circuit or spark. For the same reason the polyethylene sheet should be inspected for the presence of any holes or cracks. It is necessary that the cooling water is running before the power supply is turned on. Failure to carry out these simple precautions is the most common cause of trouble.

The Buffer

Three buffers are generally satisfactory in the separation of amino acids:

1. Formic acid 6%, pH 1.6 (682 ml of 88% formic acid in a final volume of 12 l. of water). This buffer gives reliable and reproducible results.

2. Formic acid-acetic acid buffer, pH 1.6 to 2.2 (500 ml of 85 to 90% formic acid and 600 ml of glacial acetic acid in a final volume of 10 l. of water). This buffer gives separation almost identical to the previous one.

3. Pyridine-acetic acid buffer, pH 5.4 (100 ml of pyridine and 32 ml of glacial acetic acid in a final volume of 10 l. water).

These buffers are relatively stable and can be prepared in bulk. Approximately 400 to 500 ml are poured into each vessel. Approximately the same amount of buffer should be present in each vessel to avoid the siphon effect of liquid.

Application of Specimens

Whatman® 3MM filter paper is preferable because it is thick and easy to handle when wet, and holds the disc of filter paper specimen well.

The location of sample application (origin) varies with different buffer systems. For amino acid separation at acid pH, specimens are applied near the anode. Almost all amino acids move from anode toward the cathode at pH 1.6 to 2.2. When the pH 5.4 buffer is used, the specimens are usually applied to the center of the paper. The dibasic amino acids migrate toward the cathode, and the dicarboxylic amino acids migrate toward the anode. The monoamino-monocarboxylic acids (neutral amino acids) remain near the origin.

When specimens are applied at one end and paper wicks are not used, they are spotted 5 in. from the end of the paper. When paper wicks are used, paper of shorter length is used and specimens are spotted closer to the end. For filter paper specimens, a hole is made with a paper punch at one end of the paper as far from the edge as the punch will permit. With the commercially available paper punches, the distance of the hole from the edge of the paper is usually no more than 1 in.

---

*Savant Instruments, Inc., Hicksville, N.Y. and Shandon Scientific Co., London, England

Therefore, a paper wick is necessary. The wick is made up by cutting a strip of filter paper of the same width as the paper on which samples are applied. It is folded double-layer and long enough to bridge the gap between the buffer vessel and the electrophoresis paper. The wick and the electrophoresis paper should overlap for about ¼ in. The same wick may be used for two runs. By using paper wicks, both liquid and filter paper specimens can be electrophoresed on the same sheet of paper.

Liquid urine specimens can be spotted by using disposable microcapillaries.* It is best to spot in 10 μl fractions and to dry between applications to keep the spot compact for better resolution. The specimens should be at least 1 in. apart for one-dimensional separation, and 1 ½ in. apart for two-dimensional separation. Eight specimens can be electrophoresed on 12-in. wide paper, and 12 specimens on 18-in. wide paper.

The paper is wet either by spraying or by dipping it through the buffer on both sides of the strip containing the samples. The paper is blotted between layers of filter paper (less expensive filter paper such as Whatman #1 or other types of filter paper may be used) and then the dry strip is wet with small amounts of buffer either delivered from a Pasteur pipette or by light spraying.

Operating the Electrophoresis Apparatus

When specimens are applied at one end of the paper, electrophoresis may be carried out at 2 to 3 kV(40 to 60 V/cm), 150 to 200 mA for 45 to 60 min. Half that time is adequate for specimens applied in the center of the paper. The length of time and the voltage required to give good separation depend upon several factors: the make of the apparatus, temperature of the cooling water, the wetness and size of the paper, and the ionic strength of the buffer. Optimum conditions must be determined through trial and error.

Clinical Application

Samuels[14] and Mabry and Todd[15] have reported the use of high voltage electrophoresis in the screening for amino acid abnormalities in urine. Figure 2-1 shows the abnormal patterns of several amino acid metabolic disorders by HVE. The disadvantage of using HVE alone is that too many neutral amino acids cluster in the middle, and it is very difficult to make a differential

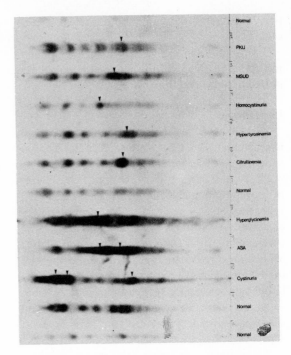

FIGURE 2-1. Urinary amino acid separation by high voltage electrophoresis at pH 1.6 (6% formic acid). Arrows indicate the abnormalities in each urine.

diagnosis without further study. In the author's opinion, for one-dimensional screening it would be simpler to do solvent paper chromatography. Probably the best is a combination of two-dimensional separation of amino acids by performing first HVE and then butanol-acetic acid-water solvent chromatography. Details of this procedure are described in the following section.

PAPER CHROMATOGRAPHY AND ITS COMBINATION WITH HIGH VOLTAGE ELECTROPHORESIS

Since the development of a simple paper chromatographic method[16] and the introduction of paper chromatography into clinical medicine[17] in the 1940's, this technique has been widely used and has proven satisfactory for screening amino acid abnormalities in blood and urine. A modification of this technique using specimens impregnated on filter paper has proved suitable for studying a large population even at a distance. Laboratories performing phenylketonuria screen-

*Microcaps®, Drummond Scientific Company, U.S.A.

ing among newborns by either the Guthrie bacterial inhibition assay or the fluorometric method can confirm positive results by paper chromatography and use the same specimen to screen for other metabolic disorders.

Equipment

Chromatography tank – Cabinets and tanks designed for ascending and descending chromatography are commercially available.

A Universal Apparatus* consists of a square glass tank, a tray for solvent, and a frame with spacers which supports five chromatograms. The chromatograms mounted on the frame can be chromatographed in one direction, dried, and turned 90° for chromatography in the second direction.

A 24 in. x 12 in. x 15 in. aquarium tank can be adapted for ascending chromatography. A stainless steel tray with handles can be used for the solvent. All the joints of the tank should be sealed with bathroom caulking material to prevent evaporation and leakage of the solvent. Molded polyethylene tanks approximately the same size as the aquarium tank are also suitable. The advantage of the polyethylene tank is that it is unbreakable and needs no tray or caulking; the disadvantage is that it is opaque.

Sewing machine – An inexpensive sewing machine with zigzag attachment may be purchased at any department store.

Hair dryer – For drying samples while spotting on chromatography paper.

Dipping trays – Different size dipping trays* designed for staining the chromatograms are available.

Chromatography oven – Large oven for drying and heating chromatograms. It is best to connect the oven with the ventilation system since most of the exhaust is irritating and has an unpleasant odor.

View box with fluorescent light – For viewing of chromatogram.

Ultraviolet light – For indoles and other fluorescent components.

Polaroid camera with stand – Optional for record keeping.

Solvents

Butanol-acetic acid-water (BuAc) (12:3:5) –

When these three components are mixed in this proportion, the solvent is completely miscible and stable over the range of temperature in the laboratory. When used in a closed tank, the solvent can be added to the tray to keep a constant level at about ½ in. deep. Emptying and cleaning of the tray should be done about once a month or when mobility of the amino acids is noticed to be slow. An overnight development is often used for the separation of amino acids. For indole separation, a short run of 5 to 6 hr is sufficient.

N-propanol-methylethylketone-25% formic acid (PrMekF) (5:3:2:)[14] – This solvent is much like the previous one in its resolution for amino acids.

Phenolic solvents[18] – The stock solution of phenol can be prepared in bulk. Add 125 ml of water to 500 g of phenol in a dark bottle with stopper. (The original bottle of phenol may be used.) A homogeneous solution can be obtained by allowing it to stand overnight or by immersing the bottle in hot running water for a few minutes. For the preparation of phenol-ammonia (PhAm) (200:1), 1 ml of concentrated ammonia is added to 200 ml of the phenol solution only as needed. This solvent may be used on two consecutive days. When ethanol is incorporated in the proportion of phenol:ethanol:ammonia:water (PhEtAm) (150:40:1:10) the solvent rises faster than PhAm, but the R_f values for amino acids are generally lower. Phenolic solvents are usually used as the second solvent in two-dimensional chromatography.

Phenol interferes with diazotized sulfanilic acid reagent (Pauly reagent); therefore, it is best not to use this solvent for imidazoles or phenolic acids that require staining with Pauly reagent. If necessary, the phenol can be washed off by dipping the paper through ether.

t-Butanol-water-methylethylketone (TBuMek) (80:80:40) – This solvent may be used for the separation of some amino acids not separated in BuAc.

Butanol-acetic acid-water (12:3:5), with 0.2% (w/v) isatin[19] – The incorporation of isatin, a location reagent, into the developing solvent is an improved method for the detection of imino acids, particularly proline. It eliminates the dipping process, increases the sensitivity, and is economical. After development is completed and the solvent allowed to evaporate, the

*Shandon Scientific Company, London, England

chromatogram is heated directly for color development. Isatin travels slower than the solvent, and a clear, colorless zone appears above the yellowish "isatin front." The difference in distance is approximately 10%. Since no compounds of interest that react with isatin have a R_f over 90, the lag of isatin does not limit the use of this method.

Isopropanol-ammonia (IPrAm) (20:1:2)[20] — The solvent is mixed as needed for indole separation. No equilibration is necessary. The mobility of indoles in this solvent may give considerable differences in absolute and relative R_f values. Therefore, it is advisable to run standards with each batch of unknown specimens. Urine can be chromatographed without desalting.

The combination of IPrAm and BuAc is recommended for general indole investigations in urine. IPrAm is used as the first solvent overnight and BuAc as the second solvent for only 5 to 6 hr (only as long as necessary for solvent to reach the top of the paper).

Potassium chloride (KCl) 20% w/v in water[20] — This solvent can be prepared in bulk. This solvent is recommended as an alternative to IPrAm for indoles. It is used as the second solvent after BuAc. The spots are compact and have constant R_f values. Caution must be taken since wet chromatograms are easily torn and they cannot be blown vigorously to dry.

Types of Specimens

For a complete study, it is recommended that both blood and urine be screened. Some disorders show marked and obvious changes in the blood amino acids, whereas others have an almost normal blood amino acid pattern, and abnormalities in the urine, on the other hand, can easily be detected. For instance, hypermethioninemia is prominent in infants with homocystinuria due to cystathionine synthase deficiency, but the urinary excretion of homocystine in the first two weeks of life may be minimal. Thus, this disorder in infants is better detected by looking for increased methionine in the blood. On the other hand, argininosuccinase deficiency, a urea cycle disorder, results in a huge amount of argininosuccinic acid (ASA) in the urine but very little in the blood. Obviously, renal transport disorders cause marked changes in urine while the blood amino acid pattern is either normal or slightly below normal. Furthermore, certain urinary amino acid changes may indicate

either an enzyme deficiency, in which there are changes in both blood and urinary amino acids (overflow or prerenal aminoaciduria), or a renal transport defect in which changes are limited to urinary amino acids. Examples of these are hyperprolinemia, due to a block in proline metabolism, and iminoglycinuria, due to a renal transport defect. Urinary findings are the same in both defects and can be differentiated only by blood proline concentration which is increased in the former and normal in the latter. If it is possible to get only one specimen, urine is preferable. More disorders can be detected with urine testing than with blood. However, confirmation of the diagnosis will, of course, require that both be studied.

CSF is the least useful for screening purposes. Changes in cerebrospinal fluid (CSF), although milder in degree, often reflect those in blood. Therefore, CSF is not recommended for screening purposes. There have been no instances where examination of CSF is critical to the diagnosis of a metabolic disorder. Only after a definitive diagnosis has been made do the abnormalities in CSF become significant, and any changes should be measured by a quantitative method.

There has been little experience with the use of amniotic fluid for antenatal diagnosis of amino acid disorders. At the present time, amniocentesis is performed in babies at risk for the few metabolic defects which can be diagnosed by specific enzyme assays in amniotic fluid cells. The performance of amniocentesis in the second trimester of pregnancy for diagnosis of amino acid disorders warrants further investigation before it can be recommended as a routine screening technique.

Collection of Specimens

There are two widely used techniques for collecting blood specimens for screening.[21,22] Whole blood can be obtained by heel or finger puncture and impregnated onto a filter paper card (Schleicher & Schuell® #903).[22] It is important that the blood be soaked through and that at least two blood spots be obtained, each about 1 cm in diameter. Blood should not be applied over the same spot repeatedly to avoid artifactual increases in amino acid concentration. After the filter paper has dried it is mailed to the laboratory.

Blood from skin puncture can be collected in a heparinized microcapillary tube, and the plasma used for analysis.[21] After the blood is collected in the microcapillary tube, one end is sealed with Plasticene. These capillary samples can then be mailed to the laboratory in a mailing container.

A random urine sample is adequate for initial screening since the changes are persistent and obvious. Thymol is usually used as a preservative, but specimens collected under chloroform, concentrated hydrochloric acid, or other preservatives may also be used. Toluene which extracts indoles should not be used if indoles are of interest.

For convenience in collection and mailing, particularly of specimens from infants, urine can be collected on the same kind of filter paper as used for blood (S. & S. #903). Urine specimens from infants can be collected by inserting a piece of specified filter paper between the layers of diaper or pressed against a wet, but unsoiled diaper. For older individuals, the filter paper is dipped into a container of urine. After the filter paper is well soaked, it is air-dried, and sent by ordinary mail. Identification should be marked with pencil on the filter paper since ink contains a number of pigments, some of which may interfere with interpretation of the results.

Preparation and the Amount of Specimens for Chromatography

Blood impregnated on filter paper must be autoclaved to prevent diffusion of blood pigment. With autoclaving for 3 min at 250°F and 15 lb of pressure no loss of amino acids has been noticed. The autoclaved specimen will be damp and should be air-dried before chromatography. For those laboratories where an autoclave is not readily available, a pressure cooker will serve the purpose. A disc either 3/16 in. in diameter containing approximately 6 μl of whole blood or 1/4 in. in diameter containing approximately 10 μl of whole blood is the appropriate amount for one-way semiquantitative chromatographic screening.

Plasma from blood collected in the microcapillary tube is separated by centrifugation in a microcapillary centrifuge. A cut segment of the tube equivalent to 5 or 10 μl of plasma can be held touching the chromatography paper to allow the plasma to flow directly on to the paper. No autoclaving is necessary.

Whole blood impregnated on filter paper may also be eluted for chromatography with a mixture of ethanol-water (60:40). A disc 12 mm in diameter, containing approximately 40 μl of blood is eluted overnight in 100 μl of the 60% ethanol. Forty microliters of the eluate are spotted for chromatography.[23] At this concentration of ethanol, satisfactory elution of amino acids takes place with no interference of the blood pigments.

For filter paper urine specimens, a 1/4 in. disc is used for one-dimensional chromatography. If the urine is very dilute, the chromatogram should be repeated with a 1/2 in. disc. For two-dimensional separation, a 3/8 in. disc is routinely used. For one-dimensional chromatography, 10 μl of liquid urine can be applied. For two-dimensional separation, the amount of creatinine in urine is used as a guideline to correct for variation in urine concentration and allow for a semiquantitative comparison of amino acid content. Urinary creatinine is estimated first by preliminary paper chromatography or by an automated chemical method (such as measurement by an autoanalyzer if available). For paper chromatographic determination of creatinine, 20 μl of the urine specimens and 20 μl each of the creatinine standards of 2.5, 5, 10, 15, 20, 30, and 40 μg/20 μl are applied to a piece of chromatography paper 6 to 7 in. wide. This is chromatographed in butanol-acetic acid-water (BuAc) for 1 to 2 hr and stained with Jaffe's reagent (see below).

Creatinine concentration in urine is estimated by visually comparing the color intensity of the creatinine spot with that of the graded standards. Urine containing 15 μg of creatinine is needed for two-dimensional separation; the volume is calculated as follows:

$$\text{volume } (\mu l) = \frac{15 \ (\mu g) \times 20 \ (\mu l)}{\text{Reading of the standard } (\mu g)}$$

Liquid urine specimens need no desalting or other preparation unless proteinuria is present. A small amount of protein does not interfere with one-way chromatography. When two-way separation of amino acids is done by a combination of high voltage electrophoresis and solvent chromatography, the urine specimen containing protein should first be deproteinized. This can be done by adding 50 mg picric acid to 2 ml urine. The mixture is then centrifuged and the clear yellow supernatant is saved. The yellow picric acid does not interfere with electrophoresis at acid pH; it moves slightly toward the anode, while most amino acids move toward the cathode.

Preparation of the Chromatogram

Whatman 3MM paper is recommended for chromatography because it gives the best resolution and holds the discs well.

For filter paper blood and urine specimens, a disc is punched out with an ordinary paper punch, and with the same punch a corresponding hole is punched on the chromatography paper. The disc of the specimen is then fitted into the hole and smoothed with a wallpaper roller.

Liquid urine or serum specimens or eluate are applied with disposable calibrated capillaries of various sizes.* The spot should be kept as compact as possible, preferably less than 1 cm in diameter. Usually 5 or 10 μl of serum are used for screening. The amount of eluate depends upon the method of elution. Ten microliter fractions are applied repeatedly and dried between applications.

For ascending chromatography, specimens are applied about 1 in. apart, along the bottom of the sheet, 1 to 1 1/2 in. from the edge (specimens should be above the solvent level). Paper of 12 in. width is suitable for an overnight run. The paper can be either rolled into a cylinder or hung in a chromatography tank. If the Universal Tank is used, corner-punched sheets of Whatman 3MM filter paper 10 in. square can be mounted to the frames. Eleven specimens can be applied to one sheet for one-dimensional chromatography. This set-up can also be used for two-dimensional separation. One sample is applied to the right lower corner of each sheet of paper. After the first run, the frame is turned 90° for the second run.

For descending chromatography, paper 18 in. wide is suitable and specimens are applied 3 in. from the edge and 3/4 in. apart. The lower end of the paper is serrated to facilitate even dripping of the solvent. The paper is hung over the trough in a chromatography tank. In general, the mobility of amino acids (R_f) is similar in both ascending and descending chromatography. Separation of some amino acids is better in descending chromatography with the longer paper than in ascending chromatography. If a descending chromatogram is allowed to develop for too long a period of time, the fast-moving compounds such as phenolic acids may run off the paper. On completion of the run, the chromatogram is hung in the fume hood or placed in an oven for drying. After drying, it is ready for staining.

Two-dimensional Separation by Solvent Chromatography

The combination of two solvent systems gives satisfactory two-dimensional separation of amino acids. Figures 2-2 and 2-3 show the amino acid pattern in a normal urine and in Hartnup disease, separated in a solvent pair of BuAc and PhAm. Other solvents and various combinations can be found in References 18, 24, and 25.

The disadvantages of two-solvent combination are that it takes two days to complete the runs and desalting of urine may be necessary.

Two-dimensional Separation, High Voltage Electrophoresis Followed by Solvent Chromatography[26]

The combination of high voltage electrophoresis using 6% formic acid and solvent chromatography using butanol-acetic acid-water (12:3:5) (BuAc) is more useful and gives better resolution than one-dimensional separation. It has been used for routine screening and also for more complete study of abnormalities found on one-dimensional chromatography.

Urine specimens are first electrophoresed as described in the previous section. Specimens are applied at least 1 1/2 in. apart. After electrophoresis, the paper is cut into 1 1/2-in.-wide strips, each containing one specimen (Figure 2-4). The strip is then sewn to another piece of Whatman

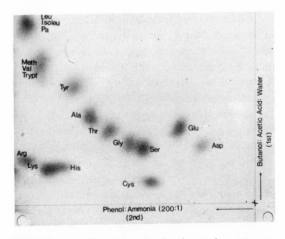

FIGURE 2-2. Two-dimensional solvent chromatogram in butanol-acetic acid-water (12:3:5) and phenol-ammonia (200:1) showing a mixture of amino acid standards.

---

*Microcaps,® Drummond Scientific Company

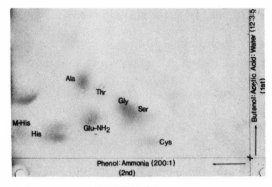

FIGURE 2-3A.

FIGURE 2-3. Two-dimensional solvent chromatograms showing (A) a normal urine and (B) urine from a patient with Hartnup disease.

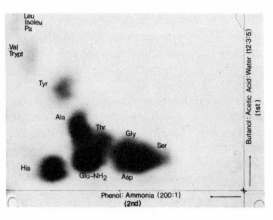

FIGURE 2-3B.

3MM paper 7 in. x 18 in. using a zigzag sewing technique, and the excess length is trimmed from each end. It is important that the paper overlap a little to give good alignment and strength. The paper is then rolled into a cylinder, ready to be put into the solvent tank. If descending chromatography is to follow electrophoresis, a 3-in.-wide wick is sewn to the other side of the electrophoresis strip.

Smith et al.[27] have adapted the two-dimensional separation by HVE and BuAc to analyze blood amino acids in filter paper specimens. They used sheets of Whatman No. 1 paper 25 x 10 cm in size, formic-acetic acid buffer pH 2, and voltage up to 400 V (16 V/cm).

When the buffer containing 1/8 of the original acid concentration is used, up to 800 V man be applied without concern for evaporation. After electrophoresis for 25 min, the paper is dried and run in BuAc for 1 hr. Only one specimen can be run on each sheet of paper.

Location Reagents[18]

After the chromatogram is dried, the location reagent can be applied either by dipping or spraying. Dipping is better than spraying in several respects.[28] It uses less reagent, takes less time, and gives more uniform distribution of the reagent than spraying. With dipping it is possible to use volatile solvents, e.g., acetone, and the colors often

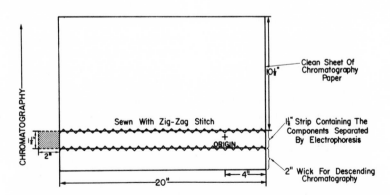

FIGURE 2-4. Diagram showing the preparation of a chromatogram for solvent chromatography following high voltage electrophoresis (HVE). For ascending chromatography, the strip containing compounds separated by HVE is sewn onto a sheet of Whatman 3MM paper by zigzag stitches. In addition, for descending chromatography, a clean paper wick is sewn onto the other side of the electrophoresis strip so there will be enough length of paper to fit in the trough and hang over the rod.

develop in a shorter period of time and are more stable. When the reagents are sprayed, the risk of inhaling toxic and irritating vapor and hazards to health are greater, and spraying should always be done in a fume hood.

Several of the commonly used location reagents are described here. Ninhydrin reagent is routinely used to stain chromatograms as a general location reagent for amino acids. One should have several other reagents on hand for the differentiation of "spots," and since compounds may overlap on the chromatogram, an abnormal pattern is not always obvious with ninhydrin color alone. A multiple dipping technique can be used or duplicate chromatograms can be prepared for special stains.

Ninhydrin reagent 0.2% (w/v) in acetone or 95% ethanol is a reagent that can be prepared and stored at 4°C in a dark bottle and will remain stable for about one month. About 10 cc of pyridine are added to 500 cc of the ninhydrin reagent before use.

The dried chromatogram is either dipped through or sprayed with the reagent. Acetone or alcohol is allowed to evaporate by placing or hanging the chromatogram in a fume hood. Color develops in several hours at room temperature. Heating of the chromatogram at 88 to 100°C for 2 to 4 min hastens the color development, especially in damp, cold weather. Colors of non-a-amino acids such as β-aminoisobutyric acid appear only after heating. Pyridine stabilizes ninhydrin color in acid atmosphere, but it decreases the intensity of fluorescence.

Most a-amino acids and compounds containing primary or secondary amino groups attached to an aliphatic carbon atom react to form purplish colors. If the chromatogram is left at room temperature, the colors will fade within several days and much faster in the presence of acid fumes. The chromatogram can be preserved for several weeks if stored at 4°C, and even longer when stored at -20°C.

Ninhydrin-cadmium acetate reagent — two stock solutions:

   a. Ninhydrin 0.25% (w/v) in acetone.
   b. Cadmium acetate 2 g in acetic acid 50 ml and water 200 ml.

The stock solutions are stored at 4°C and are mixed in proportion of 7 vol of (a) to 1 vol of (b) as needed. The dipping and color development of the chromatogram are handled in the same way as with ninhydrin reagent. Most amino acids react to form stable red to orange colors.

Isatin reagent 0.2% (w/v) in acetone is stable for at least one or two months. Before using, 2% pyridine is incorporated into the reagent.

Dip the dried chromatogram through the reagent, and after acetone has blown off or evaporated, heat the paper at 100°C for 3 to 5 min. When there is a delay between the blowing off of acetone and heating of the chromatogram, unsatisfactory color development may result. This stain is particularly useful in the location of proline and hydroxyproline. Proline, yellow with ninhydrin reagent, reacts with isatin to form a blue color. The blue color on a light yellow background is much more "visible" than yellow color on a white background. Other amino acids also react to form colors varying from blue to a reddish-purple, which are unstable and fade in several hours. The spots reappear when the chromatogram is reheated. Hydroxyproline is easily detected by applying Ehrlich reagent (see below) over isatin reagent.

Ehrlich reagent, the stock solution, p-dimethylaminobenzaldehyde 10% (w/v) in concentrated hydrochloric acid, is stable for months at room temperature. Immediately before use 1 vol of the stock solution is mixed with 4 vol of acetone. Impurities in some batches of acetone may cause a pink color, but they do not interfere with color development.

After dipping, the chromatogram should be placed flat in the fume hood to allow acetone and the acid to evaporate. Fumes from concentrated hydrochloric acid are quite irritating and may cause a film on glassware. Urea and citrulline react to form a yellow color. Hydroxyproline, citrulline, and homocitrulline can easily be detected when the ninhydrin-treated or isatin-treated chromatogram is overstained with Ehrlich reagent. Hydroxyproline appears as an immediate purple spot, whereas a stable pink color of citrulline and homocitrulline develops after the acetone has completely evaporated and the chromatogram is dry.

This reagent is useful for the location of indoles and their derivatives. The color begins to develop immediately after dipping the chromatogram. The chromatogram should be examined and the color recorded within a few hours since many of the original colors will change during that time.

Pauly reagent (diazotized sulfanilic acid reagent) — three stock solutions are made up as follows:

  a. Sulfanilic acid 1%, 9 g in 90 ml concentrated hydrochloric acid and 900 ml of water.
  b. Sodium nitrite 5%, 50 g in 1 l. water; store in dark bottle at 4°C.
  c. Sodium carbonate 10%, 100 g in 1 l. water.

As needed, add 1 vol of (a) to 1 vol of (b) in a flask. After mixing, add 2 vol of (c). Brown fumes and rising bubbles will form. This mixture should be used within 10 min. The chromatogram is dipped through the reagent and laid flat on a clean surface; colors develop immediately. Histidine, carnosine, and other imidazole derivatives react to form reddish colors. Methyl-histidines do not react. A wide range of phenolic acid derivatives gives characteristic colors with this reagent. Ammonium salt forms a yellow color. Hydroxyproline also reacts to form an orange-brown color.

Phenol vapor causes a dark brown background when the chromatogram is stained with Pauly reagent. If phenol is used in the same laboratory, chromatograms should be routinely heated at 100°C for several minutes before applying Pauly reagent.

Sakaguchi reagent — two stock solutions are prepared as follows and are stored at 4°C:

  a. Oxine 0.1% (v/v) in acetone.
  b. Bromine liquid, 0.3 ml in 100 ml of 0.5 N sodium hydroxide; discard when yellow color disappears.

Dip the chromatogram through solution (a) and allow the acetone to evaporate. Then dip it through solution (b). Arginine and other mono-substituted guanidine derivatives react to immediately form an orange color which usually fades within 30 min to several hours. Creatine and creatinine do not react with this reagent. The incorporation of urea (0.1 g is first dissolved in 0.2 ml water and then added to 100 ml of solution (a)) into oxine stabilizes the color, but this cannot be used if this chromatogram is to be dipped in Ehrlich reagent.

Cyanide-nitroprusside reagent is made up in the same manner as for the spot test of filter paper urine specimens.

First dip the dried chromatogram through 5% sodium cyanide solution in 95% ethanol.* Wait 10 to 15 min or until most of the ethanol has evaporated and the chromatogram is only slightly damp. Then dip it through the fresh alcoholic nitroprusside solution. A reddish-purple color appears immediately and fades within 10 to 15 min.

This stain is particularly useful in identifying homocystine which moves to the same location as serine in the HVE-BuAc system. Other compounds with a disulfide linkage also react to form a reddish-purple spot. Ninhydrin reagent cannot be used either before or after this stain, and a separate one-dimensional short (7 in.) chromatogram should be made for this purpose.

Iodine-azide reagent — two stock solutions are made up as follows and stored at 4°C:

  a. Iodine 0.1 N in ethanol (1.27 g in 100 ml ethanol).
  b. Sodium azide 0.5 N in 75% ethanol (3.25 g in 100 ml 75% ethanol).

Mix equal volumes of these two solutions as needed. Dip the paper and lay it flat. White spots on the brown background indicate the presence of thiols (e.g., cysteine) and disulfides (e.g., cystine, homocystine). The spots remain visible under ultraviolet light even after the brown background color fades. Thioethers (e.g., methionine) do not react with this reagent.

Platinic iodide reagent — stock solutions:

  a. Chloroplatinic acid $(H_2PtCl_6 \cdot 6H_2O)$ 1 mg/ml.
  b. Potassium iodide, 1 N (167 mg/ml).
  c. Hydrochloric acid, 2 N.

Immediately before use, mix these solutions in the following order and proportion: 5 ml of (a), 0.3 ml of (b), 0.5 ml of (c), and 25 ml of acetone. Dip the chromatogram and lay it flat for examination. As the chromatogram dries, white spots will appear on a light pink background. It may be 2 hr or more before the spots appear. Chromatography in phenol delays the reaction. Most sulfur-containing amino acids react with this

---

*It is recommended that pure ethanol be used in making reagents. It has been found that some of the compounds used to denature ethanol interfere with the reactions.

reagent, and this test is particularly useful in the detection of methionine and cystathionine.

Chromatograms stained with platinic iodide reagent may be overstained with ninhydrin.

Molybdate reagent for phosphate — the following solutions are freshly made and mixed in the proportions indicated:

a. Ammonium molybdate, 1 g in 8 ml water	4.0 ml
b. Concentrated hydrochloric acid	1.5 ml
c. Perchloric acid, 70%	2.5 ml
d. Acetone	42.0 ml

Dip the chromatogram in this mixture and allow the acetone to evaporate. Then expose it to ultraviolet light for 30 min. Phosphoethanolamine and nucleoside phosphates react to form a deep blue color. This test is useful in detecting phosphoethanolamine in the urine.

Jaffe reagent — two solutions are needed:

a. Picric acid, 1% (w/v) in 95% ethanol (stable at room temperature).

b. Potassium hydroxide, 5% (w/v) in 80% ethanol (make up fresh daily as needed).

First dip the chromatogram through the picric acid solution and allow to dry either in the cold or by heating. Then dip it through the potassium hydroxide solution and air-dry. A red color develops on a yellow background. When urine specimens are preserved in acid, creatinine appears as two red spots, running a little slower than the standards. This stain is primarily used for the determination of creatinine in urine. Creatine, arginine, and glycocyamine do not react.

Procedures in Routine Screening

The one-dimensional chromatogram of the blood is stained with ninhydrin reagent for a view of the general amino acid pattern. Figure 2-5 shows a normal blood amino acid pattern and a few examples of abnormalities. Depending upon the suspected abnormalities, it can then be over-stained with different reagents for identification of the individual amino acids which may overlap with others. As depicted in Figure 2-6, the upper 3/4 of the chromatogram is cut off below the glutamine area and overstained with Ehrlich's reagent for the detection of citrulline, hydroxyproline, and homocitrulline. Normal blood contains none of

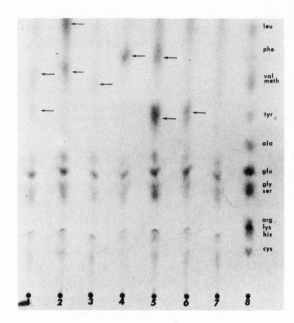

FIGURE 2-5. One-dimensional descending paper chromatogram of blood filter paper discs developed in butanol-acetic acid-water (12:3:5) showing several abnormal amino acid patterns. (1) Cystathionine synthase deficiency (homocystinuria) in a newborn; note the elevated methionine and tyrosine, the latter due to the neonatal transient hypertyrosinemia. (2) Maple syrup urine disease. (3) Hypermethioninemia due to a high protein intake. (4) Phenylketonuria. (5) Neonatal transient hypertyrosinemia with hyperphenylalaninemia. (6) Isolated hypertyrosinemia. (7) Normal blood. (8) Standard amino acid mixture. (Reproduced from Levy et al., *Clin. Biochem.*, 1, 200, 1968, with permission from the authors and publisher.)

these three amino acids in a detectable amount. The lower portion of the chromatogram is stained with Pauly reagent for histidine which usually appears as two spots in filter paper blood specimens but as one spot in serum. If the "low spots" appear intense and the Pauly stain is normal, a duplicate chromatogram should be made and stained with Sakaguchi reagent for arginine. There is no special stain for lysine and ornithine, which are both in the same location.

With 5 μl of serum or a disc 3/8 in. in diameter, the easily visible yellow color of proline is indicative of an elevation. A duplicate chromatogram can then be made and treated with isatin for a semiquantitative estimation of the proline concentration as described below.

When an elevation of amino acids such as the branched chain amino acids, methionine, phenylalanine, or tyrosine, is found, a rough estimation

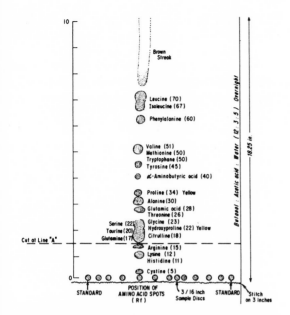

FIGURE 2-6A. Diagram of a one-dimensional chromatogram showing the position of sample discs and map of the amino acid spots in descending chromatography developed overnight in butanol-acetic acid-water (12:3:5). (Reproduced from Efron, M. L. et al., A simple chromatographic screening test for the detection of disorders of amino acid metabolism: A technique using whole blood or urine collected on filter paper, *N. Engl. J. Med.*, 270, 1378, 1964, with permission from the authors and publisher.)

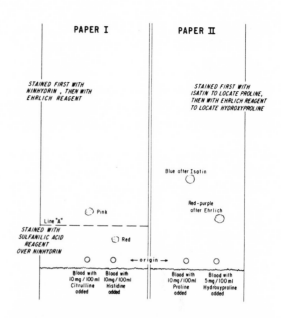

FIGURE 2-6B. Diagram of the multiple staining procedure for a one-dimensional chromatogram developed overnight in butanol-acetic acid-water (12:3:5), showing the position of abnormal spots in citrullinemia, histidinemia, and hydroxyprolinemia. (Reproduced from Efron, M. L. et al., A simple chromatographic screening test for the detection of disorders of amino acid metabolism: A technic using whole blood or urine collected on filter paper, *N. Engl. J. Med.*, 270, 1378, 1964, with permission from the authors and the publisher.)

of its concentration can be obtained by preparing another chromatogram with graded standards of the particular amino acid run in parallel, and the intensity of the spots can be compared visually with that of the standards. Because of overlapping, such estimation cannot be made for amino acids with R_f values lower than proline.

For one-dimensional chromatography of the urinary amino acids, the same procedure of cutting and multiple staining as previously described for the blood can be followed. A duplicate chromatogram is routinely prepared and developed in BuAc incorporated with 0.2% isatin for proline, and by overstaining with Ehrlich reagent, hydroxyproline, citrulline, and homocitrulline can be easily detected.

On a HVE run, the basic amino acids move the farthest, the acidic amino acids near the origin and the neutral amino acids remain in the middle. With the multiple-staining technique, the section that contains amino acids that move faster than serine is stained with Pauly or Sakaguchi reagent. The rest is stained with Ehrlich reagent.

The overlapping in two-dimensional separation is considerably less than in one-dimensional separation. A number of abnormalities are readily recognizable without multiple staining. The use of other staining techniques or additional chromatography depends upon the abnormality detected. Each of these abnormalities will be described in a later section.

THIN-LAYER CHROMATOGRAPHY

Thin-layer chromatography (TLC) is similar in principle to paper chromatography. Many of the techniques used in paper chromatography can be directly applied to TLC. Cellulose®, Avirin®, and Avicel® (the latter two materials are microcrystalline and cellulose of technical grade and pharmaceutical grade, respectively) have been found to be more satisfactory for separation of amino acids than other media, such as silica gel.

TLC requires less time and smaller quantities of specimens, and results in better separation than paper chromatography. The location reagent appears to be more sensitive because the spots are more compact. It must be applied by spraying instead of dipping. A ninhydrin solution has also been incorporated into the developing solvent. The colors developed are similar to those seen on paper and are usually more stable and last much longer. Multiple spraying is possible. Although the absolute R_f values are more variable, the relative patterns remain constant. TLC is more sensitive to salts than paper chromatography. Often desalting and concentration of urine specimens are necessary steps.

Smith et al.[29] compared chromatography of amino acids, indoles, and imidazoles on thin layers of cellulose, Avicel, and on paper and found paper chromatography to be the simplest and cheapest, and TLC on cellulose to be the fastest method. Some laboratories use TLC as the screening technique. A two-dimensional run on cellulose can be completed in a working day. This technique offers no real advantage in mass screening, because filter paper specimens can be collected and mailed, and a large number of paper chromatograms can be handled with ease. For one-dimensional paper chromatography, 20 to 25 discs of filter paper specimens can be punched out and the paper set up in a matter of minutes for overnight chromatography. Two-dimensional separation by combining HVE and paper chromatography requires less than half a working day and an overnight run. Results are obtained in less than 24 hr. For these reasons, we still prefer paper chromatography for screening.

In the following sections, a few systems of TLC for blood and urinary amino acids are described.

Equipment and Thin-layer Plates

Chromatography tanks: A variety of glass tanks are available for TLC.

Spreader: For spreading thin-layer material on the plate.

Rack: For holding plates.

Oven: For drying the plates.

Hair dryer: For drying samples between applications and plates after development.

Thin-layer plates: The thin-layer plates can be prepared in the laboratory or precoated plates can be purchased. Cellulose is best prepared as a slurry in water and ethanol (e.g., 15 to 20 g cellulose in 80 ml distilled water and 10 ml ethanol mixed in an electric mixer) and spread evenly over the entire plate with a spreader. In general, layers 0.2 to 0.5 mm thick are used. Cellulose plates are allowed to dry overnight at room temperature or heated at 80 to 90°C for 10 min. They should not be activated or heated over 90°C. The plates are reusable after cleaning. Cellulose precoated plates on disposable glass and aluminum foil are being offered by several manufacturers. One of the precoated plates* combines a strong acid cation exchanger and a cellulose layer.

The standard size of plates is 20 x 20 cm. Some prefer 10 x 10 cm or other sizes for analysis and the plate can be cut to the desired dimensions. Further details on preparation of the plates and the instrumentation in TLC are described in Reference 30.

Application of Samples

Urine is usually desalted by passing through a Dowex-50® column, eluted, and concentrated before application, and it can be desalted on a plate with an ion-exchange strip.[31] Recently, direct application of small amounts (10 μl or less) of urine and plasma has been used with success.[31-33] Blood collected on filter paper can be eluted[34] or applied directly[35] to TLC plates for chromatography. Samples are applied in 1 to 5 μl fractions and dried by a stream of warm air between applications.

Solvents and Procedures

Several solvent systems useful in separating amino acids in paper chromatography can also be used in TLC. No equilibration of the tank or use of paper liner is necessary.

Ersser and Seakins[32] reported the use of butanol-acetone-acetic acid-water (35:35:10:20) for one-dimensional separation of amino acids on precoated cellulose layers. Chromatograms are developed twice in the same solvent and dried between runs. Separation of amino acids obtained with this solvent is almost identical with that in butanol-acetic acid-water (12:3:5). Visualization of amino acids can be simplified by adding the ninhydrin solution of the solvent before the second development.[33] The plate is dried with warm air (from a hair dryer) and then heated at 80°C for 5 min for color development. The time

---

*E. M. Laboratories, Inc., Elmsford, N.Y.

required for each development is approximately 50 min.

Wadman et al.[36] described a method for two-dimensional separation on microscale chromatograms. The 20 x 20-cm precoated cellulose on aluminum foil plates is cut into 5 x 5 cm size. Urine is desalted and concentrated before application. Butanol-pyridine-water (1:1:1) is used as the first solvent and 88% phenol-25% ammonia-water (10:0.8:1) with 1 mg O-oxychinoline per 200 ml of solvent is the second solvent. The time required for the development is 30 and 50 min, respectively. Ninhydrin reagent (0.2% in ethanol) is sprayed on. Considering the small size of the chromatogram, separation of amino acids is unusually good.

Development in a combination of pyridine-acetone-58% ammonium hydroxide-water (45:30:5:20) and isopropanol-formic acid-water (150:25:25) can be completed in 4 hr. Forty amino acids found in physiological fluids have been mapped in this system by White.[37]

Precoated two-layer aluminum foil plates are now being manufactured for separation of amino acids in urine.* The lower portion (3 cm) of the plate is an ion-exchange strip and the rest is cellulose. Only 2 µl of urine are needed. The specimen is applied as a band at the left lower corner and desalted by developing the plate in water to the upper edge of the ion-exchange strip.[31] The amino acids are then moved to the first solvent front (1 cm above the ion-exchange strip) by double development in methanol: 25% aqueous ammonia (50:50). The ion-exchange strip and the portion above the second solvent front (8.5 cm above the first solvent front) are cut off. The plate is now ready for the usual two-dimensional chromatography. Pyridine:dioxane: ammonium hydroxide:water (35:35:15:15) is used as the solvent for the second run. The plate is developed twice in each of these reagents. Ninhydrin can be incorporated into the solvent before the last run. After the plate is dried, it is heated for 3 min at 80°C for color development. Overheating produces a strong background color. Figure 2-7 shows the position of various amino acids in this system. The whole operation including desalting and chromatography requires one full working day. With some modification,

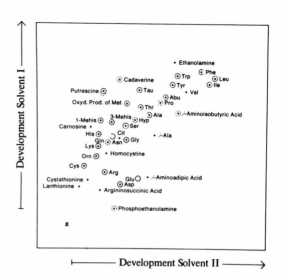

FIGURE 2-7. Diagram of a thin-layer chromatogram developed in pyridine:dioxane:ammonia:water (35:35:15:15) (solvent I) and butanol:acetone:acetic acid:water (35:35:7:32) (solvent II). Point of application is at the left lower corner. (Courtesy of Dr. Gunter Scheuerbrandt, EM Laboratories, Inc.)

urine specimens have been chromatographed in this system without desalting.[38]

Culley[35] recently described a simple technique for using filter paper specimens directly without elution. Discs 3/16 in. in diameter are placed on a TLC plate in a row about 3 cm from the bottom edge and secured in place by a glass rod. The glass rod, 8 mm in diameter and slightly longer than the plate, is held against the specimens by a heavy rubber band wrapped around both ends of the rod and across the back of the plate. The plate is then developed in ethanol:concentrated ammonium hydroxide:water (18:1:1) for about 15 min when the solvent will have migrated to about 4 cm above the spots. The plate is removed from the tank and the discs are discarded. After drying, the plate is developed in butanol-acetic acid-water (13:3:5). The slight modification of the composition of this solvent from butanol:acetic acid:water (12:3:5), according to Culley,[35] prevents the streaking of protein.

ION-EXCHANGE COLUMN CHROMATOGRAPHY

Routine quantitative analysis of amino acids using the analyzer as a screening procedure is not recommended. It is not as efficient as either paper

*E. M. Laboratories, Inc., Elmsford, N.Y.

or thin-layer chromatography. An abnormality detected by a screening paper chromatographic technique requires confirmation and further study by ion-exchange column chromatography.

Quantitative analysis by ion-exchange column chromatography is now a standard procedure. A semiautomated or automated amino acid analyzer is an instrument essential to any laboratory engaged in the study and research of amino acid metabolic disorders. Until recently, complete analysis of all the amino acids in blood and urine has been an expensive and time-consuming procedure; improved instrumentation has reduced the analysis time to 4 or 5 hr. When the diagnosis of a metabolic defect has been established and the patient is under treatment, it is not necessary to perform a complete amino acid analysis at each visit. Except for periodic examination of the whole blood aminogram to insure nutritional balance, particularly in patients on artificial diets, patients can be monitored by knowing only the levels of the involved amino acid(s) in blood using modified short column techniques. Such analysis can be accomplished in a shorter period of time and more economically than complete analysis.

Equipment

Amino acid analyzers are manufactured by a number of companies. The basic column and buffer systems and instructions for the operations of these analyzers are according to the specifications of the manufacturers. Modification is often necessary according to the needs and conditions at each laboratory. Some models are equipped with two columns, one for acidic and neutral amino acids and one for basic amino acids. Other models are equipped with a single column for all amino acids. Complete automation is possible with devices for automatic sample loading. Such devices are useful for protein hydrolysate; however, labile amino acids in physiologic fluids such as glutamine may deteriorate when samples loaded in cartridges stand at room temperature for as long as 24 hr. Some models of analyzers are equipped with a refrigerated compartment for the cartridges. The stability of these labile amino acids in the refrigerated cartridge needs to be investigated.

Types of Specimens

As a rule, fasting morning blood specimens are used for quantitative analysis. This eliminates any effect of the circadian rhythm and food intake and minimizes the variables for the sake of comparison of results between control individuals and patients and between laboratories.

A 24-hr collection of urine is usually recommended for quantitative measurement. Recently, Peters et al.[39] found the amino acid excretion measured in a timed 4-hr specimen to be comparable to that in a 24-hr collection. This shorter period of collection is more convenient. For young children and infants and where a timed specimen is not always possible, a random specimen may be used and creatinine concentration measured. The results are expressed and compared on the basis of creatinine excretion.

Preparation of Specimens and Artifacts

Both plasma and serum have been used for amino acid analysis. Plasma is preferable because clotting of the blood may cause small losses of some amino acids.[40] Venipuncture in young infants is often a time-consuming and aggravating procedure for both patient and physician. At the author's laboratory, capillary blood obtained by heel puncture has been used satisfactorily. For collection of capillary blood, a heparinized 200 μl micropipette* is filled quantitatively. The blood is emptied into a test tube containing 1 ml of 3% (or 6%) sulfosalicylic acid. The tube is mixed first by tapping and then by a mechanical mixer. In this way, the blood is deproteinized and diluted 1:5 (600 μl = 100 μl blood). This method of dilution can also be used for deproteinizing plasma when only a very small amount of specimen is available.

Heparin and EDTA (ethylenediaminetetraacetic acid) are the most commonly used anticoagulants. Excessive amounts of heparin may cause some hemolysis, leading to alteration of plasma amino acid patterns. Perry and Hansen[41] found that some batches of EDTA contain two ninhydrin-positive contaminants; one of these is eluted with taurine and the other is eluted between methionine and isoleucine in their buffer system.

Several techniques of deproteinization are avialable. The simplest procedure is to add solid sulfosalicylic acid to plasma or serum, approximately 50 mg/ml. The amount is not critical. One can weigh scoopfuls of sulfosalicylic acid using different sizes of small spatulas to find

*Microcaps®, Drummond Scientific Company

the size that holds approximately 50 mg. The spatula can be set aside for this purpose: one scoopful of sulfosalicylic acid per ml of plasma. A slight excess of sulfosalicylic acid will have no consequence. CSF contains considerably less protein than plasma so 10 to 20 mg sulfosalicylic acid per ml is adequate. Urine is also routinely prepared by adding sulfosalicylic acid 150 mg/4 ml, mainly to acidify the specimen but also to rid the urine of protein.

Picric acid is another popular deproteinizing agent; however, an extra step is necessary for its removal. One volume of plasma is added to 5 vol of 1% picric acid solution. It is mixed and centrifuged. The supernatant is passed through a small Dowex-2 column for the removal of picric acid.

Methods for removing protein from the blood have been discussed in several articles. Stein and Moore[42] reported that the picric acid method was more satisfactory than ultrafiltration or equilibrium dialysis. Gerritsen et al.[43] found that partial removal of protein by citrate buffer pH 1.5 is faster, more convenient, and more reliable than deproteinization by picric acid or sulfosalicylic acid. In their study the sulfosalicylic acid method caused losses of more than 20% of almost all amino acids except lysine and tyrosine, while the picric acid method caused some loss of at least six amino acids (aspartic acid, threonine, serine, lysine, tryptophan, and arginine). Dickinson et al.[44] found that removal of picric acid resulted in substantial losses of tryptophan and homocitrulline when compared with sulfosalicylic acid. DeWolfe et al.[45] and Perry and Hansen[41] found the sulfosalicylic acid method to be more satisfactory than the picric acid method. Our data confirm those of Hamilton[46] that picric acid and sulfosalicylic acid give identical results within the error of the method. The latter method is preferred because of its simplicity.

To avoid an artifactual change in amino acids, specimens should be deproteinized as soon as possible and stored at -20°C until analysis. Delay in deproteinization and prolonged storage even at -20°C will result in loss of cystine and homocystine which are bound to plasma proteins by the disulfide linkage. Perry and Hansen[41] observed an almost total loss of cystine after storage of the blood for 7 days at -20°C without deproteinization. Thus, the specimens should be deproteinized immediately, centrifuged, and the supernatant separated for storage. Storage at -68°C is more satisfactory than storage at -20°C.[44] Glutamine decreases and glutamic acid increases during storage. If these are the amino acids of particular interest, specimens should be analyzed shortly after collection.

Blood cells contain large amounts of glutamic acid, aspartic acid, phosphoethanolamine, and glutathione, and very little cystine, methionine, and arginine.[47] Arginase, which hydrolyzes arginine to ornithine and urea, is present in significant quantity in blood cells. "Leakage" of this enzyme from the cells results in a decrease of arginine and an increase in ornithine in the plasma. Thus, alterations in the concentrations of these amino acids can be expected in a hemolyzed blood specimen.

The amount of specimen to be analyzed varies with the sensitivity of the analyzer. Roughly 0.1 to 0.5 ml blood and 0.25 to 1.0 ml CSF are appropriate amounts. For a timed urine specimen, a 30-sec volume or less is used; if preliminary paper chromatography or TLC shows hyperaminoaciduria, the amount should be decreased accordingly. For instance, in a patient with Hartnup disease as little as 5 or 10 μl are enough. If the amount to be applied is very small, it is advisable to dilute the specimen 1 to 10 in order to measure a reasonably accurate amount.

Quantitative measurement of argininosuccinic acid (ASA) on the analyzer is complicated by the instability of free ASA which has a tendency to form cyclic anhydrides in acidic solution. It is best to convert all ASA to its stable anhydrides before analysis by first acidifying the urine specimen to pH 2 and then placing it in a boiling water bath for 2 1/2 hr.[48] Often only 10 or 25 μl are adequate for the measurement of ASA. For other amino acids in the same specimen, particularly glutamine, a separate analysis with an unboiled fresh specimen must be made.

For the total amount of free and bound amino acids in urine, the specimen is hydrolyzed in 6 N hydrochloric acid at 125°C for 3 hr (equal amounts of urine and concentrated hydrochloric acid in a sealed tube). The hydrolysate is desalted on a Dowex-50 column. The eluate is then evaporated to dryness using a rotary evaporator and reconstituted to a known volume with 0.1 N hydrochloric acid or the starting buffer for analysis.

Buffers

Buffers for analysis can be purchased in ready-to-use solutions or, more economically, prepared in the laboratory. Deionized water and an accurate pH meter are essential. When satisfactory separation is obtained with one batch of buffer, it might be advantageous to save a small portion of that particular batch and titrate the new buffer to the same pH (even if the absolute pH is slightly different from the specified one) to insure good and reproducible results.

The standard procedure for amino acid analysis employs sodium citrate buffer. The disadvantage of the sodium buffer is that it gives poor separation of glutamine, asparagine, and serine. This presents no problem for protein hydrolysate, but may present problems in physiological fluids. Recently, it has been found that substitution of lithium citrate buffer for sodium citrate buffer improves the separation of these amino acids.[49-52] The disadvantage of the lithium buffer[49] is that the column is packed more tightly, building up pressure and causing some loss of resolution. Frequent repouring of the column is then necessary. This problem can be minimized by operating the column at low temperature for the minimum amount of time, regenerating it at 70°, and avoiding extreme change of ionic strength of the buffer by limiting the lithium molarity of the final buffer to less than 1.2. Application of samples in sucrose solution should be avoided. It is also helpful to insert a Teflon® disc on top of the resin column and pack another 2 to 3 cm resin on top of the disc (Figure 2-8). This top portion of the resin can be easily removed for cleaning when it collects particles and becomes discolored from the samples and from the air-pressure line. Drying out of the column or blowing air into it causes a rise in pressure and poor resolution and should definitely be avoided. When these precautions are taken while using any kind of buffer, it will save the extra work involved in frequent repouring of the column.

Operation of Columns

Complete Analysis

The method described by Moore and Stein[53] uses a two-column and a discontinuous buffer system. The acidic and neutral amino acids are eluted from a long column and the basic amino acids from a short column. The method described by Piez and Morris[54] uses a single column and a gradient change of buffers for the elution of all amino acids. Other methods are mainly variations and modifications of these two basic systems.

For physiologic fluid it is necessary to start with a low operating temperature (33 to 37°C) to separate serine and glutamine and to ensure accurate measurement of glutamine which is unstable and easily cyclizes to form a ninhydrin-negative compound at higher temperature. In addition, the separation of glutamic acid, proline, and citrulline, and α-aminobutyric acid is affected by temperature. In the buffer system described by Efron,[55] resolution of these amino acids on a 100 cm column can only be achieved by using a 37°C temperature for the first 2 1/2 hr. Perry et al.[49] found that 35°C for the first 6 1/2 hr is necessary in their system using lithium buffer. The length of time for "low" temperature varies with the exact length of the resin column, the flow rate, and the buffer system, and should be guided by the manufacturer's suggestions and methods described in the literature. Individual laboratories can work out the conditions most satisfactory for them.

Elution of certain amino acids is sensitive to the change of pH. For instance, the position of cystine

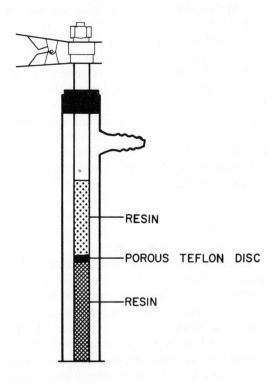

FIGURE 2-8. Diagram showing a column of two resins separated by a Teflon® disc; the top resin can easily be removed for cleaning.

is greatly affected by pH, but not much by temperature change. Ionic strength of the buffer is another factor which affects the elution of some amino acids more than others. Readers should refer to the articles by Moore and Stein,[53,56] Hamilton,[57] and Zacharius and Talley[58] for detailed discussion of these and also for mapping of over 100 compounds in their systems.

The flow rate of the buffer is adjusted according to the size and length of the column, efficiency of the pump, and the resolution of the peaks.

Several nonphysiologic amino acids which are eluted from the column at a time when they do not interfere with any known compounds in physiologic fluids have been used as an internal standard. Norleucine was widely used in the past. However, in our buffer system, as well as in others, it is eluted with cysteinehomocysteine disulfide, a compound present in blood and urine of patients with homocystinuria and has thus been abandoned. Homocitrulline, used by Gerritsen et al.,[43] is not a suitable amino acid since this compound is excreted by normal infants and by patients with hyperornithinemia (Table 1-1). We have now resorted to the use of cysteic acid as a standard. This compound is eluted before sample application, partly as a check to insure that the ninhydrin and recording systems are in good working order.

At our laboratory a one-column analyzer (Technicon) is modified in such a way that two analyses can be done simultaneously instead of the manufacturer's recommended one analysis per operation. The analyzer is equipped with three colorimeters; two of these read at 570 nm (one is half as sensitive as the other). The third colorimeter reads at 440 nm for the detection of proline, hydroxyproline, and other compounds which are ninhydrin-yellow. Since the channel with the half-sensitivity at 750 nm is rarely useful, this colorimeter is altered by using a cuvette with a longer light path to give the same sensitivity as the other 570 nm channel. The eluate from a second column is mixed with ninhydrin reagent, reacted in a separate coil, and fed into this channel. For urine analysis in many metabolic disorders the omission of readings at 440 nm presents no real problem. Scriver et al.[59] have similarly modified their two-column analyzer (Beckman-Spinco) for simultaneous analysis of two samples. The effluent from the long column for the neutral and acidic amino acids and that from the short column for the basic amino acids are analyzed at the same time so that a complete analysis is achieved in a considerably shorter period of time.

Examples of blood amino acid analysis using a one-column analyzer and a gradient change of buffers are given in Figures 2-9 and 2-10.

Selected Amino Acids

Short column chromatography for the rapid separation of selected amino acids in physiologic fluids has been developed by several groups.[59–62] Most of the amino acids of interest can be eluted within 2 hr. Resin columns of 18 to 29 cm are used. It is much more efficient if two columns are set up. A sample can be eluted from one column while the other column is being regenerated. Therefore, continuous analysis can be performed.

In addition to the short column method, the operation of the column and the analyzer can be modified in such a way that a group of compounds can be detected by their elution pattern and their specific colorimeter reactions.

Acidic amino acids that are eluted at the very beginning of the run, such as phosphoethanolamine, cysteic acid, cysteine-sulfinic acid, and taurine, are not well separated with the standard buffer system. DeMarco et al.[63,64] described a system using $0.1\,M$ citric acid as the eluent buffer for the resolution of these amino acids. Anion-exchange resin is necessary for more acidic compounds.

Argininosuccinic acid and its anhydrides can be further identified and separated from other amino acids by elution from a short column of anion-exchange resin according to the method described by Ratner and Kunkemueller.[65]

Sulfur amino acids can be eluted from the column and identified simultaneously by automated color reactions with ninhydrin reagent, nitroprusside reagent, and iodoplatinate reagent using a modified manifold of an amino acid analyzer.[66]

Imidazole derivatives can be studied by elution from the same column used in amino acid analysis using sodium citrate buffers and diazotized sulfanilic acid reagent for colorimetric analysis of the column eluent.[67] Since phenolic acids in urine interfere with the color reaction, they should be removed by extraction with ethylacetate.

A review of the analytical procedure for the determination of amino acids in blood and urine by ion-exchange chromatography has recently been published.[68]

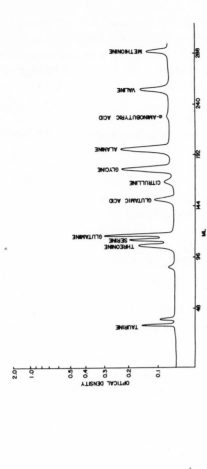

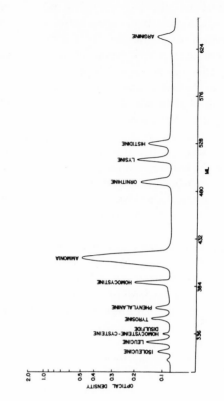

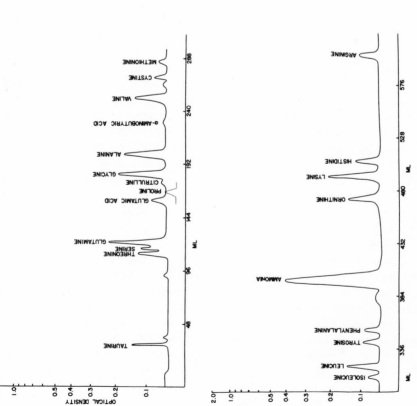

FIGURE 2-9. Amino acid analysis of a normal plasma using a Technicon amino acid analyzer Model MC-1. Two hundred microliters of the deproteinized plasma were applied to a 0.9 x 50 cm column. A gradient buffer system was used for elution. The buffer was contained in a nine-chambered device (Autograd). Seventy-five milliliters of 0.05 *M* sodium citrate buffer, pH 2.875 were in chambers 1 to 5; 40 ml of this pH 2.875 buffer and 35 ml of an 0.267 *M* sodium citrate buffer, pH 5.00 in sixth chamber; 75 ml of the pH 5.0 buffer were in chambers 7 to 9. The flow rate was 48 ml/hr. The column temperature was at 37°C for 1 hr and 45 min and increased to 60°C for the remainder of the analysis. Total analysis time was 13 1/2 hr.

FIGURE 2-10. Amino acid analysis of plasma specimen from a patient with cystathionine synthase deficiency (homocystinuria) showing elevated methionine and the presence of homocystine. Analyzer condition as described in Figure 2-9

Clinical Application of Ion-exchange Chromatography

The whole amino acid pattern should be analyzed when confirming an abnormality detected by a screening test or when initially studying a patient. Certain disorders are impossible to diagnose by paper chromatography alone; for instance, hyperglycinemia is often unnoticeable on paper. Hyperornithinemia and hyperpipecolatemia cannot be diagnosed without quantitative measurement of blood amino acids on the analyzer. When hyperammonemia is found, a complete analysis of the blood and urine amino acid patterns should be performed on an analyzer because the relationship between ammonia metabolism and the metabolism of other amino acids is still not fully understood.

It is noteworthy that when an elevation detected by paper chromatography of a filter paper blood specimen cannot be confirmed by analysis of plasma on the analyzer, whole blood should be analyzed. Whole blood can be prepared by diluting with 6% sulfosalicylic acid as for capillary blood. An uneven distribution of arginine between the red cells and plasma is known to occur in crab-eater monkeys with erythrocytes lacking arginase activity.[67] A similar situation has never been observed in humans. It is possible that the general use of plasma instead of whole blood for amino acid analysis precludes its discovery.

The short column method is the most suitable for monitoring dietary therapy in several metabolic disorders, especially if the amino acid concentration is kept below 5 mg/100 ml (300 to 400 μmol/l.), a level not reliable with semiquantitative paper chromatography or bacterial inhibition assay, and for certain amino acids that cannot be resolved on paper. Figures 2-11 to 2-21 illustrate the different systems useful in the study of various disorders as described by Shih et al.[60,62] and give examples of abnormalities seen in these conditions.

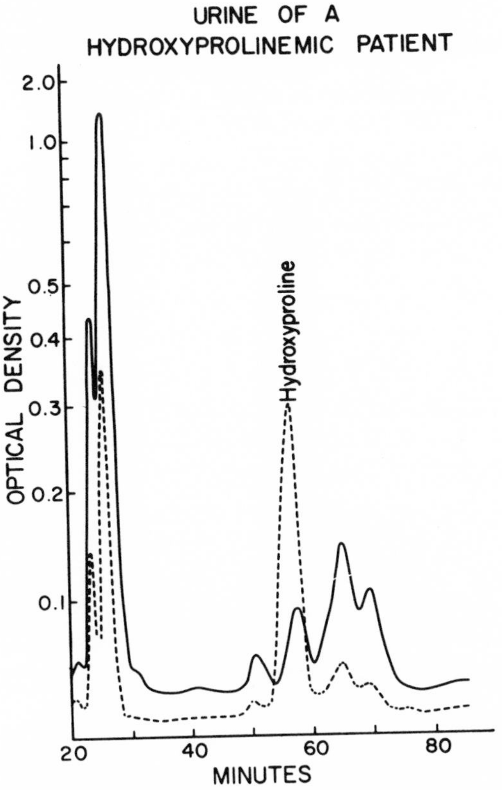

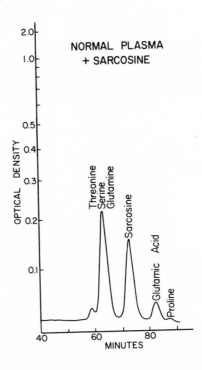

URINE OF A HYDROXYPROLINEMIC PATIENT

FIGURE 2-11. Separation of hydroxyproline on a 0.9 x 20 cm column using a pH 2.45, 0.2 *M* sodium citrate buffer, 1 ml/min at 33°C. (Reproduced from Shih, V. E. et al., Rapid short-column chromatography of amino acids: A method for blood and urine specimens in the diagnosis and treatment of metabolic disease, *Anal. Biochem.*, 20, 299, 1967, with permission from the authors and the publisher.)

FIGURE 2-12. Separation of sarcosine on a 0.9 x 20 cm column using a pH 2.65, 0.2 *M* sodium citrate buffer, 1 ml/min at 65°C. (Reproduced from Shih, V. E. et al., Rapid short-column chromatography of amino acids: A method for blood and urine specimens in the diagnosis and treatment of metabolic disease, *Anal. Biochem.*, 20, 299, 1967, with permission from the authors and the publisher.)

33

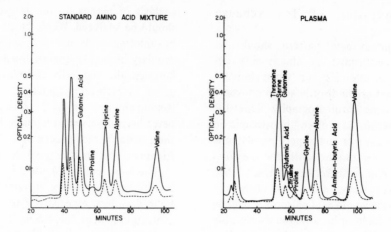

FIGURE 2-13.   Separation of proline, glycine, alanine and α-aminobutyric acid on a 0.9 x 20 cm column using a pH 2.875, 0.2 *M* sodium citrate buffer, 1 ml/min at 65°C. (Reproduced from Shih, V. E. et al., Rapid short-column chromatography of amino acids: A method for blood and urine specimens in the diagnosis and treatment of metabolic disease, *Anal. Biochem.,* 20, 299, 1967, with permission from the authors and the publisher.)

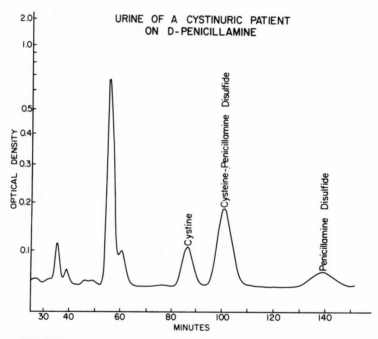

FIGURE 2-14.   Short column (0.9 x 20 cm) separation of amino acids in urine from a cystinuric patient on penicillamine therapy, using a pH 3.0, 0.2 *M* sodium citrate buffer, 1 ml/min at 65°C. (Reproduced from Shih, V. E. et al., Rapid short-column chromatography of amino acids: A method for blood and urine specimens in the diagnosis and treatment of metabolic disease, *Anal. Biochem.,* 20, 299, 1967, with permission from the authors and the publisher.)

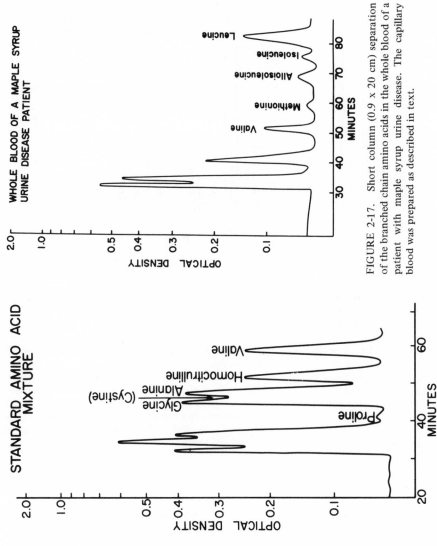

FIGURE 2-17. Short column (0.9 x 20 cm) separation of the branched chain amino acids in the whole blood of a patient with maple syrup urine disease. The capillary blood was prepared as described in text.

FIGURE 2-16. Separation of homocitrulline on a 0.9 x 20 cm column, using a pH 3.30, 0.22 M sodium citrate buffer 1 ml/min at 65°C. (Reproduced from Shih, V. E. et al., Rapid short-column chromatography of amino acids: A method for blood and urine specimens in the diagnosis and treatment of metabolic disease, *Anal. Biochem.*, 20, 299, 1967, with permission from the authors and the publisher.)

FIGURE 2-15. Separation of methionine and cystathionine on a 0.9 x 20 cm column, using a pH 3.25, 0.2 M sodium citrate buffer, 1 ml/min at 65°C. (Reproduced from Shih, V. E. et al., Rapid short-column chromatography of amino acids: A method for blood and urine specimens in the diagnosis and treatment of metabolic disease, *Anal. Biochem.*, 20, 299, 1967, with permission from the authors and the publisher.)

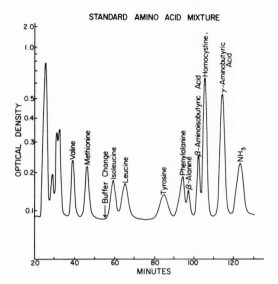

FIGURE 2-18. A short column (0.9 x 20 cm) system suitable for study of homocystinuria and hyper-betaalaninemia, using a pH 3.50, 0.24 M sodium citrate buffer until after the emergence of methionine, followed by pH 4.30, 0.4 M sodium citrate buffer. (Reproduced from Shih, V. E. et al., Rapid short-column chromatography of amino acids: A method for blood and urine specimens in the diagnosis and treatment of metabolic disease, *Anal. Biochem.*, 20, 299, 1967, with permission from the authors and the publisher.)

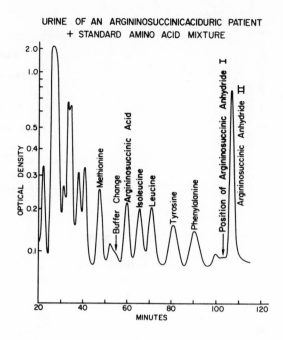

FIGURE 2-19. Separation of argininosuccinic acid and its anhydrides on a 0.9 x 20 cm column using the same system as described in Figure 2-18. (Reproduced from Shih, V. E. et al., Rapid short-column chromatography of amino acids: A method for blood and urine specimens in the diagnosis and treatment of metabolic disease, *Anal. Biochem.*, 20, 299, 1967, with permission from the authors and the publisher.)

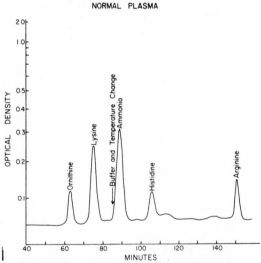

FIGURE 2-20. Separation of basic amino acids on a 0.9 x 20 cm column, using a pH 4.79, 0.4 M sodium citrate buffer 1 ml/min at 33°C until after the emergence of ammonia, at which time the buffer is changed to pH 5.8, 0.8 M sodium citrate and temperature to 65°C. (Reproduced from Shih, V. E. et. al., Rapid short-column chromatography of amino acids: A method for blood and urine specimens in the diagnosis and treatment of metabolic disease, *Anal. Biochem.*, 20, 299, 1967, with permission from the authors and the publisher.)

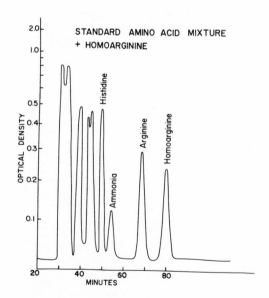

FIGURE 2-21. Separation of arginine and homoarginine on a 0.9 x 20 cm column using a pH 5.2, 0.9 M sodium citrate buffer, 1 ml/min at 65°C.

Normal and Abnormal Amino Acid Patterns

Table 2-2 lists the data from several studies of blood amino acid values in normal healthy individuals measured by ion-exchange column chromatography. With few exceptions, the amino acid concentrations in blood are remarkably constant in different age groups, races, sexes, and people with different dietary habits. In general, children in the active growth stage have slightly lower concentrations of the essential amino acids. There are no significant differences in amino acid concentrations between older people 52 to 95 years of age and younger adults.[75]

Amino acid concentrations in whole blood obtained by heel puncture (Table 2-2) are in general similar to those in plasma, although higher values of glutamic acid, aspartic acid, and ornithine, and lower values of cystine and methionine can be expected in the whole blood.

Diets with ordinary protein contents (2 to 3.5 g/kg/day) have no significant effect on plasma amino acid levels in infants. Severe protein malnutrition in children (kwashiorkor) causes a decrease in the essential amino acids and a reversed ratio of essential to nonessential amino acids. Young infants fed whole milk may have a protein intake as high as 7 to 10 g/kg/day. In those infants many of the plasma amino acids, primarily the essential amino acids, are increased two to three times normal. Methionine elevation is most striking — three to ten times normal.[70,76]

Urine is a complex physiologic fluid; its amino acid content varies with age, diet, and other factors, as will be discussed in a later section. Hamilton[77] found that it contained at least 175 ninhydrin-positive compounds using a high resolution column. It is, therefore, conceivable that one peak that elutes from an ordinary column contains two or three compounds.

For identification of an unknown amino acid, the ratio of the optical absorbance at 570 nm and 440 nm which is constant for a given compound, and the R_f values of the unknown in at least five solvents should be compared with that of the authentic compound and proven to be identical.

Normal values of urinary amino acids are listed in Table 2-3. It is difficult to collect data on

normal infants and no large series has been reported.

Cerebrospinal fluid contains small quantities of amino acids; most of the amino acids are present in concentrations of only 5 to 15% of those found in plasma, except glutamine, asparagine, serine, and threonine which are close to 1/2 their concentrations in plasma, and proline which is only about 5%.[85-87] Table 2-4 lists the amino acid values in CSF as reported by various investigators. Cystathionine and γ-aminobutyric acid (GABA) which occur in the brain in large quantity have not been found in CSF in normal conditions or in brain injury.

In the majority of disorders the abnormal amino acid is present in higher concentrations in blood than in CSF. The exceptions are argininosuccinic aciduria and aspartylglycosaminuria where the accumulation of the abnormal metabolite is greater in CSF than in blood.

Large amounts of bound or conjugated amino acids are excreted in the urine. These bound amino acids are liberated by acid hydrolysis. The increase in amino acids over the amounts before hydrolysis is a measurement of the bound amino acids. A major portion of the bound amino acids in urine is composed of glycine, glutamic acid, and aspartic acid.[39,82] Considerable amounts of proline, cystine, threonine, serine, valine, tyrosine, lysine, and hydroxyproline are also excreted. The amount of bound amino acids (Table 2-5) in urine is quite variable and may be affected by diet. Increased excretion has been found in active bone disease and severe burns. Total bound hydroxyproline excretion is increased in a number of disease states such as Marfan syndrome, burn, and hyperparathyroidism.*

GAS-LIQUID CHROMATOGRAPHY

The usefulness of gas-liquid chromatography (GLC) in amino acid analysis is limited. Amino acids are nonvolatile compounds and must be converted to volatile derivatives before GLC application. It has been difficult to find a derivatization technique or single chromatographic system suitable for the analysis of more than 175

---

*Bound hydroxyproline should not be confused with free hydroxyproline. The bound hydroxyproline but not the free hydroxyproline is increased in the diseases mentioned above. On the other hand, only the free form is increased in hydroxyprolinemia, an inborn error of hydroxyproline metabolism (hydroxyproline oxidase deficiency).

TABLE 2-2

Blood Amino Acid Concentrations (μmoles/1.)[a] in Different Age Groups

Age	Prematures (1 day) 10 Dickinson et al.[69]		Newborns (1 day) 25 Dickinson et al.[44]		Infants (1 to 3 months) 29 Snyderman et al.[70]		Infants (2 to 6 months 7 Shih		Children (9 month 2 years) 20 Soupart[71]	
Number of subjects Reference										
Amino acid	Mean[d]	±SD	Mean	Range	Mean	±SD	Mean	Range	Mean	Range
Taurine	180	75	141	74–216	25	14	61	46–86	49	19–91
Hydroxyproline	40	40	32	–	–	–	–	–	–	–
Aspartic acid	10	10	8	TR–16	4	2	447	247–687	2	0–9
Asparagine	–	–	45	–	31	8	–	–	135	46–29
Threonine	215	60	217	114–335	144	40	260	191–364	–	–
Serine	270[e]	75	164	94–243	114	19	276[e]	196–467	60	33–12
Glutamine and asparagine	905[f]	250	764	538–959	–	–	336[f]	232–450	92	24–17
Proline	230	75	185	107–277	201	62	287	128–597	115	51–18
Glutamic acid	65	35	52	20–107	–	–	234	174–328	–	–
Citrulline	–	–	16	9–29	21	11	38	31–50	–	–
Glycine	460	275	344	224–514	164	29	243	175–296	170	56–30
Alanine	375	50	330	236–410	275	71	300	177–413	219	99–31
AAB[c]	–	–	13	6–29	16	9	28	TR–38	–	–
Valine	130	50	137	80–246	194	49	295	160–553	127	57–26
Half cystine	65	10	61	36–84	54	21	66	53–81	–	–
Methionine	35	5	29	9–41	21	9	34	16–49	29	3–29
Isoleucine	40	20	40	27–53	59	14	80	38–123	44	26–94
Leucine	70	25	72	47–109	104	30	153	80–229	75	45–15
Tyrosine	120	100	70	42–99	82	26	133	72–216	45	11–12
Phenylalanine	90	20	79	42–110	52	14	80	52–97	40	23–69
Ornithine	90	20	92	49–151	72	23	147	104–215	40	10–10
Lysine	190	60	200	114–269	103	33	220	135–341	87	45–14
Histidine	50	20	77	49–114	63	10	110	96–137	64	24–11
Arginine	50	20	54	22–88	48	13	81	56–142	31	11–65
Tryptophan	30	15	32	TR–67	–	–	–	–	–	–

[a]All measurements were performed in either plasma or serum except that of Shih, where capilliary whole blood, prepared as descri in text, was used; when values were reported in mg they were recalculated to μmoles.

[b]Data from nine laboratories compiled by Dickinson et al.[44]

[c]AAB = a-aminobutyric acid

[d]As adapted from original article by Scriver et al.[74]

[e]Includes asparagine

[f]Glutamine alone

TABLE 2-2 (continued)

Blood Amino Acid Concentrations (μmoles/l.)[a] in Different Age Groups

Children (3 to 10 years) 9 Scriver and Davies[72]		Children (5 to 17 years) 39 Nyhan et al.[73]		Adults 8 Dickinson et al.[44]		76[b]		Adults 10 Efron and Shih	
Mean	Range	Mean	±SD	Mean	Range	Mean	Range	Mean	±SD
80	57−115	69	6	66	46−138	63	27−168	108	59
−	−	−	−	−	−	−	−	−	−
−	25	29	4	16	TR-53	7	0−24	−	−
10	4−20	−	−	43	−	44	41−49	−	−
76	42−95	119	4	163	102−246	129	79−193	108	42
94	79−112	157	6	112	65−193	115	73−167	120	42
−	−	376	27	568	415−694	568	−	559	113
106	68−148	149	8	235	111−446	184	102−336	153	112
110	23−250	186	16	58	17−118	58	14−192	58	19
−	−	−	−	30	12−55	29	−	29	12
166	117−223	242	20	232	144−488	237	120−554	199	63
234	137−305	338	12	345	249−502	336	210−661	286	124
−	−	−	−	16	10−23	20	8−35	22	8
162	128−283	190	6	170	116−227	213	141−317	182	31
60	45−77	−	−	74	48−140	44	9−84	74	29
14	11−16	22	2	21	15−26	23	6−40	19	5
43	28−84	61	2	54	35−88	63	37−98	54	11
85	56−178	11	4	101	71−136	111	75−175	108	20
43	31−71	57	2	50	36−86	72	22−87	49	18
42	26−61	61	3	58	38−116	53	37−88	49	8
33	27−86	59	5	70	33−126	60	30−106	89	18
111	71−151	167	8	174	144−211	153	83−238	224	43
55	24−85	88	3	80	63−93	74	32−107	124	12
53	23−86	96	8	82	49−151	75	21−138	76	33
−	−	−	−	48	25−73	−	−	−	−

TABLE 2-3

Urinary Free Amino Acids in Different Age Groups

Age	Newborn (1 day)	Infants (10 days to 7 weeks)		Children (3 to 12 yrs)		Adults		Adults		Adults		Adults		Adults	
Number of subjects	10	5		12		20		8		10		10		10	
Reference	Armstrong et al.[78]	Levy et al.[79]		Carver and Paska[80]		Scott-Emuakpor et al.[81]		Stein[82]		Logan et al.[83]		Berridge et al.[84]		Shih	
	Mean	Range		Range		Range		Range		Range		Mean[a] ± SD		Mean[b] ± SD	
Amino acid	μm/mg creatinine	mg/24°	μm/24°	mg/24°	μm/24°	mg/24°	μm/24°	mg/24°	μm/24°	mg/24°	μm/24°	μg/mg creatinine		μg/mg creatinine	
Taurine	3.73	3.0–20	26–157	7.9–121.2	63.1–969	—	—	86–294	687–2349	10–38	78–304	55.3	41	110	31
Hydroxyproline	.19	—	—	—	—	—	—	—	—	—	—	—	—	0	—
Aspartic acid	TR	—	—	<5	<37	TR-27	TR-197	<10	<548	0.5–8	4–57	2.2[e]	.8	—	—
Threonine	.21	1.0–12	13–100	10–30	85–249	22–16	187–1370	15–53	126–392	6–24	54–202	13.7	7.2	27[f]	25
Serine	.55	6–25	59–235	16–57	155–540	14–126	129–1387	27–73[c]	257–695	18–81	172–772	23.3	12.3	46[d,e]	26
Glutamine	.34	12–26	85–177	21–114	140–779	—	—	34–92[c]	257–696	—	—	40	18.5	46[e]	18
Asparagine	—	—	—	—	—	—	—	—	—	—	—	6.4	2.4	—	—
Proline	.11	3–11	28–96	0	—	0	—	<10	<87	—	—	—	—	0	—
Glutamic acid	.07	0.3–2	2–10	0	—	TR-34	TR-230	<10	<68	2–10	11–68	3.5	1.1	7[g]	5
Citrulline	—	TR	TR	0	—	0-TR	0-TR	—	—	0.4–8	2–43	1.8[f]	.8	4	—
Glycine	1.19	15–59	194–787	12–107	165–1420	59–294	785–3918	68–99	906–1319	13–135	169–1798	59.6	23.9	104	50
Alanine	.30	4–10	46–104	9–39	102–439	8–48	88–541	21–71	236–797	2–37	20–415	13.7	5.5	24	11
α-aminobutyric acid	—	TR	TR	2–8	16–77	—	—	—	—	—	—	1.0[j]	.43	<2	—
Valine	.08	1.0–3.0	12–27	2–6	15–51	TR-7	TR-57	<10	—	2–12	21–102	4.2	1.2	8[g]	2
1/2 Cystine	.09	2.2–3.3	18–28	5–31	41–257	TR-38	TR-317	<10–21	<83–175	8–26	67–216	7.6	3.2	9	3
Homocitrulline	0.3	0.8–2.0	1–13	—	—	TR-11	TR-58	—	—	—	—	—	—	—	—
Methionine	TR-.04	TR-.4	TR-3	3–14	20–95	TR-9	TR-62	<10	—	0.3–3	2–17	4.5	2.5	2[i]	1
Isoleucine	.23	0.9–2.0	7–15	2–7	18–56	TR-8	TR-58	12–28	92–214	0.7–3	5–24	2.9	1.1	3[j]	1
Leucine	.06	—	—	3–11	21–83	TR-11	TR-86	9–24	69–183	3–8	20–62	3.8[g]	1.7	5	1
Tyrosine	.04	4–7	22–40	7–31	40–168	TR-42	TR-229	15–49	83–270	8–27	44–149	11.3	6.0	17	4

TABLE 2-3 (continued)

Urinary Free Amino Acids in Different Age Groups

Phenylalanine	.04	1–2	7–10	4–18	24–106	TR-17	TR-100	9–31	55–188	3–17	18–103	5.7	2.1	9	2
β-aminoisobutyric acid	.18	0.2–0.8	2–7	–	–	–	–	–	–	–	–	–	–	–	–
Hydroxylysines	.06	–	–	<5.0	<38	TR-7	TR-53	–	–	0.6–13	5–102	1.3[h]	.4	18[g]	4
Ornithine	.02	2–3	17–21	–	–	–	–	–	–	–	–	–	–	–	–
Lysine	.31	6–11	39–75	9–94	64–642	3–153	21–1048	7–48	48–328	8–53	53–365	7.6[g]	2.1	83	63
1-methylhistidine	.18	0.8–2	5–10	4–46	24–273	–	–	–	–	–	–	25.7[g]	12.1	–	–
Histidine	.24	16–39	103–249	48–199	306–1285	73–441	470–2843	113–320	728–2062	50–196	322–1263	71.2	34.9	120	73
3-methylhistidine	.10	1–4	7–21	9–42	56–2485	–	–	–	50–196	–	–	27.3	6.3	33	7
Tryptophan	–	–	–	–	–	–	–	–	–	3–245	–	–	–	–	–
Carnosine	.08	–	–	–	–	–	–	–	–	–	–	–	–	–	–
Arginine	.04	TR-1	TR-7	<5	<29	TR-50	TR-288	<10	<57	0.4–5	2–26	2.2	.9	4[g]	1

All values are taken from original reports and calculated to include both mg and μmol when values are expressed on a 24-hour basis.

[a] Mean of 11 observations unless otherwise indicated.
[b] Mean of ten observations unless otherwise indicated.
[c] Serine and asparagine together, values calculated as one half each.
[d] Serine and asparagine.
[e] Mean of seven observations.
[f] Mean of six observations.
[g] Mean of nine observations.
[h] Mean of ten observations.
[i] Mean of five observations.
[j] Mean of eight observations.

TABLE 2-4

Amino Acid Concentrations in Cerebrospinal Fluid (micromoles/100 ml)

Reference Amino acid	Perry et al.[85] Mean ± SEM*	Dickinson and Hamilton[86] Mean ± SD**	Van Sande et al.[87] Mean ± SD***
Taurine	0.64 ± 0.05	0.63 ± 0.18	0.53 ± 0.14
Aspartic acid	–	0.09 ± 0.05	0.29 ± 0.27
Threonine	3.15 ± 0.15	2.48 ± 1.01	2.66 ± 0.93
Serine	2.35 ± 0.12	3.78 ± 2.29****	3.57 ± 0.96****
Asparagine	0.74 ± 0.03	–	–
Glutamine	58.97 ± 3.39	50.9 ± 14.4	45.47 ± 12.08
Glutamic acid	0.17 ± 0.03	0.70 ± 0.49	1.47 ± 1.33
Proline	–	0.06 ± 0.16	Trace
Citrulline	0.27 ± 0.02	0.20 ± 0.08	0.21 ± 0.07
Glycine	0.58 ± 0.03	0.66 ± 0.18	0.85 ± 0.25
Alanine	3.27 ± 0.21	2.32 ± 0.94	2.79 ± 0.99
α-Aminobutyric acid	0.38 ± 0.06	0.34 ± 0.19	0.26 ± 0.08
Valine	2.09 ± 0.16	1.46 ± 0.55	1.43 ± 0.40
Cystine	–	0.02 ± 0.03	Trace
Methioine	0.25 ± 0.02	0.26 ± 0.16	0.32 ± 0.10
Isoleucine	0.53 ± 0.05	0.44 ± 0.13	0.50 ± 0.09
Leucine	1.49 ± 0.11	1.09 ± 0.36	1.16 ± 0.24
Tyrosine	0.90 ± 0.08	0.91 ± 0.50	0.79 ± 0.23
Phenylalanine	0.95 ± 0.07	0.92 ± 0.58	0.75 ± 0.22
Ethanolamine	1.21 ± 0.15	1.28 ± 0.57	0.67 ± 0.33
Ornithine	0.49 ± 0.06	0.57 ± 0.18	0.84 ± 0.23
Lysine	2.91 ± 0.16	1.87 ± 0.66	1.86 ± 0.64
Histidine	1.20 ± 0.05	1.30 ± 0.44	1.11 ± 0.29
Homocarnosine	–	–	0.27 ± 0.12
Arginine	2.16 ± 0.13	2.01 ± 0.58	1.42 ± 0.74

Abbreviations
SEM = Standard errors of mean
SD = Standard deviation
*Ten adults
**Specimens collected for neurological diagnostic procedures from 8 females and 10 males, age from 4 months to 42 years.
***Specimens from 13 patients with no significant neurological disorders.
****Serine and asparagine together.

TABLE 2-5

Bound Amino Acids in Adult Urine

Number Reference Amino acid	3 Stein[82] Range mg/24°	5[a] Berridge et al.[84] Mean ± SD (μg/mg creatinine)
Taurine	0 – 13	2.8 ± 2.2[b]
Aspartic acid	192 – 251	87.1 ± 24.1
Threonine	34 – 45	14.1 ± 3.6
Serine	37 – 82	15.4 ± 4.2
Glutamic acid	470 – 640	200 ±58
Proline	67 – 94	43.6 ± 12.2[c]
Glycine	680 – 940	145 ± 66
Alanine	24 – 45	15.9 ± 3.8
Cystine	42 – 52	–
Valine	22 – 40	12.2 ± 4.2
Isoleucine	3 – 8	3.7 ± 1.2
Leucine	5 – 14	9.4 ± 3.8
Tyrosine	20 – 62	3.2 ± 1.4[c]
Phenylalanine	10 – 32	4.8 ± 2.2
Histidine	33 – 130	15.3 ± 7.5[d]
Methylhistidine	-1 – 10	–
Lysine	37 – 60	26.6 ± 19.2
Arginine	–	11.8 ± 2.6

[a]Mean of ten observations in five subjects
[b]Mean of four observations
[c]Mean of eight observations
[d]Mean of nine observations

amino acids in physiologic fluids. The derivatization procedure is laborious and time-consuming. Although analysis of the derivatives by GLC is fast and efficient, the time involved in preparing the derivatives is substantial and the total work effort is no less than is required for analysis by ion-exchange chromatography. With further improvement in methodology and when used in combination with mass spectrometry, GLC will become a useful technique in the study of amino acid abnormalities (see Chapter 4 for more discussion).

MICROBIOLOGICAL ASSAYS OF AMINO ACIDS

Before the advent of ion-exchange chromatography, especially the automated analyzer method, quantitative measurement of natural amino acids was usually carried out by tube microbiological assays. The technique utilizing microorganisms has now been modified to a simple and practical method for mass screening for inborn errors of metabolism. Dried blood specimens on filter paper are used for all these tests, whether it is for detecting an abnormal amount of amino acids or an absence of an enzyme.

In 1963 Guthrie and Susi[88] reported a plate method of bacterial inhibition assay for phenylalanine concentration in blood specimens collected on filter paper. This simple and economical assay has since been introduced in many states and countries as a screening test for PKU in newborns and has proved to be valuable and practical. Tests for the elevation of several other compounds or enzyme defects other than PKU were subsequently devised by the same group.*[89] These bacterial inhibition assays and metabolite inhibition assays are often informally referred to as "Guthrie tests."

The plate assay methods for eight amino acids (lysine, methionine, cystine, threonine, tryptophan, tyrosine, arginine, and histidine) using lactic acid bacteria, as described by Bolinder,[90] were developed primarily for studying the nutritional values of protein and have not been applied to clinical medicine in the study of inborn errors of metabolism. The basic principles of these plate assay methods are similar to that of the "Guthrie tests," but liquid samples are used instead of dried filter paper samples. Details of the methods will not be given here and interested readers should refer to the comprehensive treatise by Bolinder.[90]

Principles

Bacterial inhibition assays are simple agar diffusion assay techniques based upon the inhibition of normal bacterial growth by an antimetabolite of the amino acid to be tested in a minimal culture medium and the reversal of such inhibition by the presence of the test amino acid. Small filter paper discs impregnated with blood or urine are placed upon the surface of the agar culture medium. After overnight incubation, the diameter of a growth zone surrounding the disc is a measure of the concentration of the test compound in the specimen. Control discs with known amounts of the test compound are placed on each agar plate for comparison.

Bacterial metabolite inhibition assay differs from the bacterial inhibition assay in that the test compounds specifically inhibit the growth of the test organism. Only one such test has been developed for the detection of galactosemia and hypervalinemia.

Some of the enzymes in dried blood on filter paper remain stable for long periods of time and the measurement of their activity can be used as a screening test for enzyme defect. The test for galactose-1-phosphate uridyltransferase deficiency (galactosemia) developed by Beutler and Baluda (see Chapter 3) is an example. Recently, Murphey et al.[91] devised other such tests utilizing the growth response of a mutant bacteria when a nonutilizable substrate is converted by the enzyme into a growth-promoting product. The absence of a growth zone around a blood disc indicates the lack of the test enzyme in the specimen.

Bacterial Inhibition Assays for Amino Acids

Bacterial inhibition assays for phenylalanine, methionine, leucine, tyrosine, histidine, and lysine are now available.[89-91] Table 2-6 lists the genetic disorders for which the assays are devised and the organisms and inhibitors used.

---

*Dr. Guthrie's laboratory acts as a ready reference for up-to-date information of the techniques. For any laboratory with the intention of setting up these microbiological assays it is advisable to contact Dr. Robert Guthrie, Children's Hospital, Buffalo, New York for helpful hints.

TABLE 2-6

"Guthrie Bacterial Inhibition Assays" for Amino Acids as Screening Tests

Test amino acid	Disorder	Organism	Inhibitor (Concentration in test medium)
Phenylalanine	Phenylketonuria	*B. subtilis* ATCC 6633	β-2-Thienylalanine $(1.3 \times 10^{-6}\ M)$
Leucine	Maple syrup urine disease	*B. subtilis* ATCC 6051	4-Azaleucine $(6.7 \times 10^{-5}\ M)$
Methionine	Cystathionine synthase deficiency (homocystinuria)	*B. subtilis* ATCC 6633	Methionine sulfoximine $(1.5 \times 10^{-6}\ M)$
Tyrosine	Tyrosinosis (hereditary tyrosinemia)	*B. subtilis* ATCC 6051	d-Tyrosine $(1.5 \times 10^{-6}\ M)$
Histidine	Histidinemia	*B. subtilis* ATCC 6051	1,2,4-Triazole-3-alanine $(6 \times 10^{-6}\ M)$
Lysine	Hyperlysinemia	*B. subtilis* ATCC 6051	S(β-Aminoethyl) cysteine $(5 \times 10^{-5}\ M)$

Equipment

Paper punch, 1/8 in. in diameter.

Styrene plastic trays,* 7 in. x 11 in. with cover for test agar plate.

Punching machine (for automation),** a machine for automatic numbering, punching, and placing four discs of blood specimens simultaneously on four separate agar plates in a predetermined pattern.

Incubator.

Preparation for Assay Medium

The formula is a modified Demain's minimal medium for *B. subtilis* spore germination. The composition of the medium is as follows:

	g /l.
Dextrose	10.0
K_2HPO_4	30.0
KH_2PO_4	10.0
NH_4Cl	5.0
NH_4NO_3	1.0
Na_2SO_4	1.0
Glutamic acid	1.0
Asparagine	1.0
L-alanine	0.5
Salt solution	10.0 ml

(a mixture of
$MgSO_4 \cdot 7H_2O$ 60.0 g, $MnCl_2 \cdot 4H_2O$ 1.0 g,
$FeCl_3 \cdot 6H_2O$ 1.0 g, and $CaCl_2$ 0.5 g in 1 l.)

Dextrose is prepared as a 10% solution and sterilized separately. The remainder of the contents is dissolved in 900 ml distilled water and the pH will be 6.8 to 7.0. This solution is poured into bottles of convenient sizes and sterilized by autoclaving. Before use, 9 vol of the solution are mixed with 1 vol of 10% dextrose solution. Dehydrated culture medium is commercially available.***

Preparation of Inhibitors

The test response is a function of the concentration of inhibitor and the reaction of the bacteria. Concentrations of the inhibitors as listed in Table 2-6 give the optimum growth zone in the range of 1 to 10 mg/100 ml of the test amino acid. Solutions of the inhibitors are prepared at such a concentration that less than 0.5 ml of the solution per plate is needed to achieve the desired concentration. The optimum amount of inhibitor may vary a little from the recommended dose, depending upon the response of the different strains of *B. subtilis*.

Preparation of the Inoculum

Spores of *B. subtilis* ATCC 6633 and 6051 are

*Bufkor Inc., Buffalo, N.Y.

**Developed by Strauss, Phillips, Valberg, and Guthrie and manufactured by Fundamental Products Company, North Hollywood, Calif.

***Baltimore Biological Laboratories (BBL), Baltimore, Md.; Ames Company, Elkhart, Ind.

supplied in 10 ml vials by commercial source.* B. subtilis ATCC 6633 is prepared by diluting 1 vol to 10 vol with Demain's medium, B. subtilis ATCC 6051 1 vol to 5 vol. One tenth milliliter of the preparation is used per plate.

Preparation of the Agar Test Plate

For the preparation of 3% agar, 30 g of agar are dissolved in distilled water by placing the bottle in a boiling water bath with gentle shaking. After the solution is brought to boil and the agar completely dissolved, it is cooled to 50 to 55°C.

Warm the Demain's medium to 50 to 55°C, add the proper amounts of inhibitor and the inoculum, and mix. One equal volume of the warm agar solution is slowly poured into the bottle of medium and mixed thoroughly. Then 150 ml of the mixture are dispensed into each individual tray, 7 in. x 11 in., to give a layer of proper thickness. A thicker layer reduces the intensity of growth zone. The trays are placed on a level table and allowed to harden. The trays are reusable; they should be washed clean, but no sterilization is necessary.

Collection and Preparation of Specimens

Blood is obtained from a heel puncture and impregnated on filter paper, Schleicher and Schuell #903 (the same type for paper chromatography). Four spots, each at least 3/8 in. in diameter, are collected. Handling of the specimens is the same as for those collected for paper chromatography. In fact, the same specimen can be used for both types of test. Specimens are autoclaved for 3 min and air-dried before applying to the plates.

Preparation of Controls

Control blood specimens containing known amounts of the test amino acid can be either purchased or prepared at individual laboratories. Outdated bank blood or animal blood (e.g.,

horse blood) is obtained and assayed for amino acid contents by an amino acid analyzer or other convenient procedure. An amount of the test amino acid is then added to the blood to make concentrations 2, 4, 6, 8, 10, 12, and 20 mg/100 ml. The blood is applied with a pipette to a sheet of S.S. #903 filter paper to make spots of 3/8 to 1/2 in. diameter. After the blood spots are air-dried, the sheets are stored dessicated at 4°C. These control specimens should be autoclaved as with the unknown.

Procedure

After the agar is hardened, the plates are ready for use. Discs of specimens 1/8 in. in diameter can be punched by hand and placed with forceps in rows on an agar test plate 1 in. apart. When handling a larger number of specimens, this step can be automated by using a punching machine** A series of control specimens is placed in the middle row. Each plate will accommodate 72 discs in 6 rows of 12. Sixty unknown specimens can be tested on each plate. The trays are incubated overnight at 37°C.

Results and Clinical Applications

The agar plates are examined with the aid of both transmitted and direct light. The halo of growth zone around each unknown blood disc is compared with that around the control discs (Figure 2-22). The levels at which a follow-up study is required have been determined arbitrarily and by experience, and the following cut-off points are currently in practice: phenylalanine and leucine, 4 mg/100 ml; methionine, 2 mg/100 ml; tyrosine, 8 mg/100 ml; and histidine, 6 mg/100 ml. A "positive" result should always be confirmed by repeating the bacterial inhibition assay to ensure the correct sample identification, and by analysis using another method such as chromatography or fluorometric method.*** Paper

*Baltimore Biological Laboratory, Baltimore, Md.; Ames Company, Elkhart, Ind.
**Fundamental Products Company, North Hollywood, Calif.
***The fluorometric method for the determination of phenylalanine was first described by McCaman and Robins:[93] serum or plasma is deproteinized by adding an equal volume of 0.6 N trichloroacetic acid. Ten microliters of the supernatant solution are mixed with 100 μl succinate buffer (0.3 M, pH 5.8), 50 μl ninhydrin (30 mM), and 20 μl L-leucylalanine in a test tube and incubated at 60°C in a water bath for 120 min. After cooling, add 1 ml of copper reagent (1.6 g sodium carbonate, 65 mg potassium sodium tartrate, and 60 mg copper sulfate ($CuSO_4 \cdot 5H_2O$) in 1 l. water). The tubes are read after 10 to 15 min in a fluorometer with a primary filter of 365 nm and secondary filters of 505 to 530 nm. Fluorescence intensity of the unknowns is compared with that of phenylalanine standards. This method has been adapted to automation and to the use of filter paper specimens by Hill et al.[94] and by Ambrose.[95] With the automated procedure, phenylalanine in 60 samples can be measured in 1 hr. Fluorometric methods for the determination of tyrosine[96,97] and of histidine[98] have also been described.

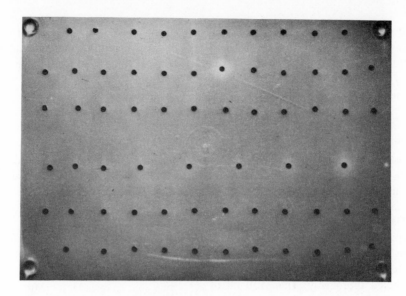

FIGURE 2-22. Bacterial inhibition assay. Discs in the fourth row are control specimens with known amounts of test amino acid (2, 4, 6, 8, 12, and 20 mg/100 ml). Other discs are unknown blood specimens. The specimen in the seventh position of the second row shows a growth zone comparable to the control specimen of 20 mg/100 ml.

chromatography will reveal the whole amino acid pattern and detect other coexisting abnormalities. For column chromatography (amino acid analyzer) blood is eluted from the filter paper and the eluate applied to the column.[92] Rapid analysis by short column methods[60,62] is extremely useful.

Contamination of the culture medium or the sample and an insufficient amount of inhibitor will result in overgrowth of bacteria, obscuring the reading of results. No false-negative results have been reported. Blood specimens taken from patients receiving antibiotics do not inhibit the bacterial growth response to phenylalanine.[99] Occasionally there is a complete lack of growth around the disc which is easily distinguishable from a normal growth response. Complete inhibition indicates the presence of an inhibitor in the sample, and another blood specimen should be obtained for retesting. These bacterial inhibition assays are more sensitive than paper chromatography in detecting mild elevations and are particularly useful in newborn screening. The reliability of the bacterial inhibition assay for phenylalanine has been evaluated[100] and it compares favorably with the McCaman-Robins fluorometric method. There is now little doubt that the former is a simple and effective test for the early diagnosis of PKU.[101,102] Furthermore, with the set-up of bacterial inhibition assay, other

amino acids such as methionine and leucine can be included in screening with little extra work or expense. Of the six inhibition assays (Table 2-6), all except the one for lysine have been field-tested by various laboratories engaged in newborn screening and have been found effective in detecting elevations.

Since all these tests are based upon elevations of the affected amino acids, the timing of blood sampling is very important. The blood sample should not be collected until after the infant has had at least 12 milk feedings.

Metabolite Inhibition Assay

An *E. coli* mutant lacking galactose-1-phosphate uridyltransferase activity is inhibited by galactose, galactose-1-phosphate, and valine. In the presence of these compounds there is an inhibition zone rather than growth zone surrounding the blood disc. Therefore, this test should detect the two types of galactosemia (Table 1-3) as well as hypervalinemia. Further discussion of this test can be found in Chapter 3.

Microbiological Screening Test for Enzyme Deficiency

A test for argininosuccinase activity[91] is the only available microbiological screening test for an enzyme deficiency in amino acid metabolism. The same filter paper blood specimen for routine screening is used. Since this test is for enzyme activity, a portion of the specimen should be cut off and set aside before autoclaving.

Equipment

Same equipment as for the inhibition assays is required.

Preparation of Assay Medium

Demain's medium is prepared in the same fashion as described previously.

Preparation of Inoculum

Spores of *Bacillus subtilis* JP3, an arginine mutant derived from *B. subtilis* ATCC 6051, are harvested from the surface of an arginine-containing medium* and stored in a refrigerator. A spore suspension of .05 to 0.2 ml, having an optical density of 0.9 at 550 nm, is used per plate (150 ml medium).

Preparation of Argininosuccinate

The barium salt of argininosuccinic acid (ASA) is available commercially.** As needed, 0.1 M sodium argininosuccinate is prepared from the barium salt.*** Barium argininosuccinate**** is first dissolved in water, and an equimolar amount of a cold sodium sulfate solution is added. Barium sulfate precipitate is removed by centrifugation at 4°C. The pH of the solution should be adjusted to neutral or between 7 and 8, as argininosuccinic acid tends to form the inactive cyclic anhydrides at acid pH. This preparation should be stored frozen and used within one week. Each lot of the barium salt of ASA should be tested for the occasional presence of growth-promoting contaminant. Contamination of the ASA will result in growth of the mutant without the presence of arginine. On the other hand, lack of growth around control blood discs suggests that free ASA may have formed cyclic anhydrides and become inactive. In either case, a fresh solution of ASA should be prepared.

Procedure

Blood discs 1/8 in. in diameter are placed on the agar plate and the plate is incubated overnight at 37°C.

Results and Clinical Application

A growth zone around the disc of blood indicates argininosuccinase activity and there is no growth around the disc lacking this enzyme activity. Occasional blood specimens contain inhibitors and no growth is observed. These specimens also produce zones of inhibition on plates containing 10^{-4} arginine. A "positive" (no growth) result should therefore be confirmed in a repeat blood specimen and by detection of ASA in blood and urine.

Since a malignant type (neonatal variant) of argininosuccinic aciduria may cause neonatal death, it is advisable to do the screening test as early as possible. This enzyme test, which does not depend upon the accumulation of metabolites, can be performed on cord blood if available.

INTERPRETATION AND FURTHER INVESTIGATION OF ABNORMALITIES

Normal Amino Acid Patterns and Their Physiological Variations in Blood and Urine

Concentrations of blood and cerebrospinal fluid amino acids are relatively constant (Figures 2-5, 2-23, and 2-24). Samples taken in the afternoon usually contain more amino acids because of the presence of circadian rhythm. When young infants are fed a high protein diet, generalized hyper-aminoacidemia and hyperaminoaciduria[70,76] may be observed. Methionine may be greatly elevated out of proportion to the other amino acids. Therefore, it should be cautioned that the finding of hypermethioninemia in an infant does not always indicate cystathionine synthase deficiency, a disorder often known as homocystinuria.

Tyrosine elevation is the most frequent finding in newborn screening. Transient neonatal hyper-tyrosinemia, the most common cause of tyrosine elevation, occurs in approximately 0.5 to 1.8% of full-term infants and in as many as 1/3 of premature infants. This type of transient hyper-tyrosinemia responds to large doses of ascorbic

---

*Potato homogenate 10% w/v, glucose 0.01%, tryptone 0.01%, yeast extract 0.01%, agar 1.5%, and arginine 10^{-4} M.

**Sigma Company, St. Louis, Mo. and Calbiochem, San Diego, Calif.

***This compound should be handled very carefully. It is extremely hydropic and should be stored dessicated at 0 to 5°C. Even during weighing, it is advisable to keep the container in the dessicator.

****Because of variable amounts of barium precipitated with argininosuccinate, the weight equivalent of 1 μmol of each lot may be different and should be obtained from the supplier.

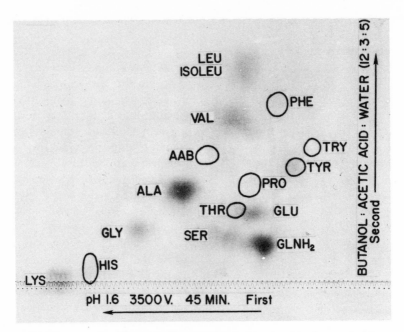

FIGURE 2-23. Two-dimensional paper chromatogram of 1 ml normal plasma. High voltage electrophoresis (HVE) at pH 1.6 (6% formic acid) was followed by overnight ascending chromatography in butanol-acetic acid-water (BuAc) (12:3:5). (Reproduced from Efron, M. L., High voltage paper electrophoresis, in *Chromatographic and Electrophoretic Techniques, Volume II, Zone Electrophoresis,* 3rd ed., Smith, I., Ed., John Wiley & Sons, New York, 1968, 166, with permission from the author and publisher).

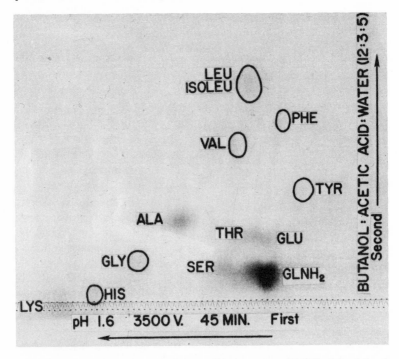

FIGURE 2-24. Two-dimensional paper chromatogram (HVE-BuAc) of 1 ml cerebrospinal fluid. (Reproduced from Efron, M. L., High voltage paper electrophoresis, in *Chromatographic and Electrophoretic Techniques, Volume II, Zone Electrophoresis,* 3rd ed., Smith, I., Ed., John Wiley & Sons, New York, 1968, 166, with permission from the author and publisher.)

acid. Normalization of tyrosine excretion after oral administration of 100 to 200 mg ascorbic acid daily for several days usually indicates that this is the benign form of hypertyrosinemia. Hyperphenylalaninemia may be secondary to the accumulation of tyrosine. It is imperative that hypertyrosinemia be ruled out in those infants with moderate phenylalanine elevation so that a diagnosis of phenylalanine metabolic disorder, PKU, will not be erroneously made. The diagnosis of PKU is highly improbable when hyperphenylalaninemia and hypertyrosinemia coexist.

Age is an important factor in the amino acid excretion pattern. Urinary amino acid excretion is rather constant in adulthood (Figure 2-25), about 150 to 200 mg amino acid nitrogen in 24 hr. Infants under six months of age excrete relatively more amino acids than do older children and adults. They have a characteristic "baby pattern" which consists of the increased excretion of proline, hydroxyproline, and glycine (Figure 2-26). Sometimes cystine and lysine are also excreted in increased quantity in infants, making the diagnosis of heterozygocity for cystinuria more difficult in early infancy.

Taurine excretion is marked at birth and decreases rapidly to an undetectable level during the first week of life. Infants who are breast-fed continue to excrete taurine. A small quantity of carnosine is a normal constituent of the urine of adults and infants. Transient mild generalized aminoaciduria is not uncommon in patients hospitalized with acute illnesses of various types. During pregnancy histidine excretion is increased. β-Aminoisobutyric aciduria is a genetic marker, being more common in Orientals and people of Mediterranean origin than in Caucasians.

Abnormal Amino Acid Patterns

An abnormal pattern is easy to recognize when there are 10 or 25 specimens on 1 sheet of a one-dimensional chromatogram. The abnormality will stand out as different from the others. Figures 2-5 and 2-27 show the different patterns in various amino acid metabolic disorders on one-dimensional chromatograms developed in BuAc. They can be compared with Figure 2-1 which depicts the abnormal urinary patterns as seen after HVE at pH 1.6. Two-dimensional chromatograms illustrating these abnormalities are shown below. For comparison, the amino acids found in physiologic fluids are mapped in Figure 2-28.

When any abnormality is suspected from the initial specimen, it has been our policy to obtain a repeat specimen before reaching a definite conclusion. When an unusual pattern is seen, the

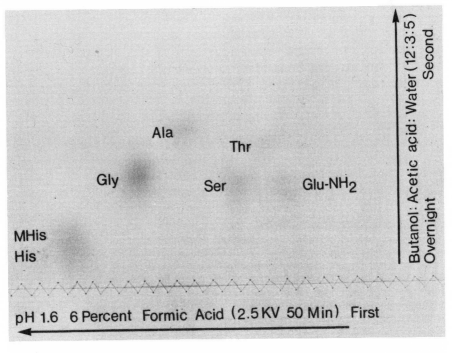

FIGURE 2-25. Two-dimensional paper chromatogram (HVE-BuAc) showing the normal urinary amino acid pattern.

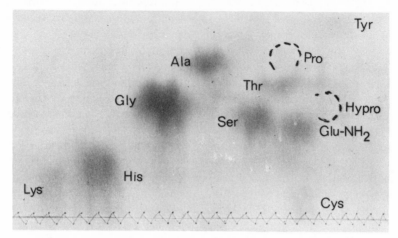

FIGURE 2-26. Two-dimensional paper chromatogram (HVE-BuAc) showing the normal amino acid pattern in the urine of an infant. Note the increased amount of proline, hydroxyproline, glycine, lysine, and cystine ("baby pattern").

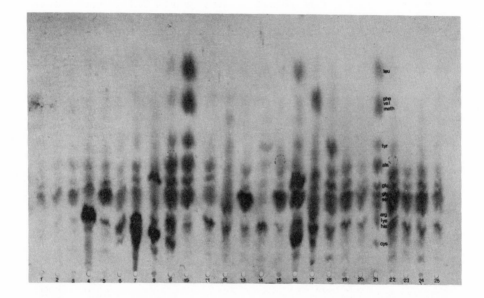

FIGURE 2-27. One-dimensional descending chromatogram of filter paper urine discs developed overnight in butanol-acetic acid-water (12:3:5) and stained with ninhydrin reagent. Dotted line outlines proline. 1, 2, 3. normal; 4. argininosuccinic aciduria; 5. citrullinemia; 6. heterozygous cystinuria; 7. homozygous cystinuria; 8. cystathioninuria; 9. generalized hyperaminoaciduria (Fanconi syndrome); 10. Hartnup disease; 11. histidinemia; 12. homocystinuria and hypermethioninuria; 13. hyperglycinemia; 14. hyperlysinemia; 15. iminoglycinuria; 16. maple syrup urine disease; 17. phenylketonuria; 18. hypertyrosinemia; 19, 20. normal; 21. amino acid standards; 22. normal infant showing a prominent "baby pattern" (increased amounts of tyrosine, proline, glycine, hydroxyproline, lysine, and cystine); 23. normal; 24. normal infant showing a moderate "baby pattern" (increased amounts of proline, hydroxyproline, and glycine); 25. normal.

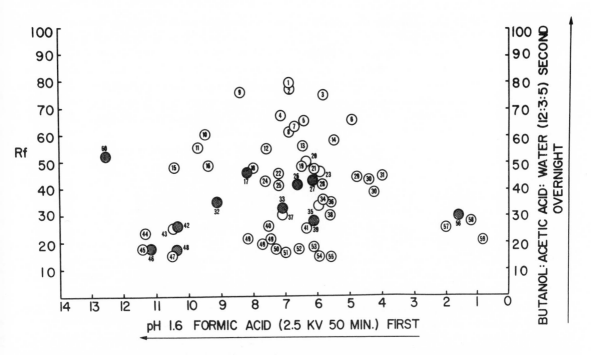

FIGURE 2-28. Map of amino acids separated by high voltage electrophoresis at pH 1.6 and followed by overnight development in butanol-acetic acid-water (12:3:5). Shaded circles indicate those amino acids present in normal urines. All amino acids react to form purple spots with ninhydrin 0.2% in ethanol except as indicated: (1) Leucine; (2) Isoleucine; (3) Phenylalanine; (4) Valine; (5) Methionine; (6) Tryptophan (brownish purple); (7) Kynurenine; (8) Pipecolic acid; (9) Tyramine (grayish purple); (10) β-aminoisobutyric acid; (11) γ-aminobutyric acid; (12) α-aminobutyric acid; (13) 3-hydroxykynurenine (brown); (14) Tyrosine; (15) β-alanine (blue); (16) δ-aminolevulinic acid (brownish yellow); (17) Alanine; (18) Glycyl-proline (yellow); (19) Proline (yellow); (20) Prolylhydroxyproline (yellow); (21) α-aminoadipic acid; (22) α-acetyl ornithine; (23) δ-acetyl-ornithine; (24) Sarcosine; (25) Hydroxyprolylglycine; (26) Threonine; (27) Glutamic acid; (28) Homocitrulline; (29) 3,4-dihydroxy-phenylalanine (DOPA); (30) β-mercaptolactate-cysteine disulfide; (31) 5-hydroxytryptophan; (32) Glycine; (33) Serine and homocystine; (34) Aspartic acid; (35) Citrulline; (36) Hydroxyproline (brown); (37) S-adenosyl-homocysteine; (38) γ-hydroxyglutamic acid; (39) Glutamine; (40) S-adenosyl-methionine; (41) Asparagine; (42) Methylhistidines; (43) Arginine; (44) Carnosine; anserine (brown); (45) Ornithine; (46) Lysine; (47) Hydroxylysine; (48) Histidine (grayish purple); (49) Argininosuccinic acid and its anhydrides; (50) Homolanthionine; (51) Cystathionine; (52) Homocysteine-cysteine disulfide; (53) Saccharopine; (54) Cystine; (55) Glutathione (oxidized); (56) Taurine; (57) Phosphoethanolamine; (58) Cysteine sulfinic acid; (59) Cysteic acid; and (60) Ethanolamine.

chromatogram is repeated with the same specimen in order to be certain that the unusual pattern is not an artifact of the preparation and staining process and that there has been no mix-up of specimens. Certain metabolic diseases such as MSUD, ketotic hyperglycinemia, and tyrosinosis, which may be life-threatening but will respond to early treatment, should be recognized and reported immediately.

Not all abnormal amino acid patterns are due to endogenous or hereditary metabolic errors. Urinary excretion of ninhydrin-reactive compounds is affected by a number of factors including diet and drug ingestion (exogenous aminoaciduria). When an abnormality is seen, the possibility of exogenous origin should first be ruled out. For this reason, exogenous or secondary aminoaciduria will be discussed first.

We have found that Polaroid® pictures of unusual chromatograms can be extremely useful both for record keeping and for later comparison.

Exogenous or Secondary Aminoaciduria
Aminoaciduria of Dietary Origin

The excretion of several amino acids is affected by diet. Homocitrulline is present in the urine of normal infants who are fed canned formula or milk. It is pathologic only when it is found in older children not on canned milk and/or in combination with hyperornithinemia, hyperammonemia, hyperlysinemia, or other amino acid abnormalities.

Almost all infant soy formulae are supplemented with DL-methionine.* Most of the D-methionine, which is poorly reabsorbed by renal tubules and is utilized very little by the body, is excreted in the urine.[103,104] Endogenous methioninuria consists predominantly of the L-form and can be differentiated by taking a dietary history and employing a specific bacterial inhibition assay to measure L-methionine only. A preparation of DL-methionine,** which is marketed commercially as an oral medication for diaper rash, can also cause D-methioninuria. In dietary methioninuria, the blood methionine is within normal range and the cyanide-nitroprusside test is usually negative. These two negative findings readily rule out homocystinuria due to cystathionine synthase deficiency. Methionine malabsorption syndrome can be ruled out by reexamination of the urinary amino acids after discontinuation of D-methionine-containing formula or medication.

The ingestion of a large amount of poultry, particularly white meat, often results in the excretion of carnosine, anserine, and 1-methylhistidine. The consumption of certain plant foods causes urinary excretion of phenolic and indole amines.[105] Ingestion of collagen-containing food, e.g., gelatin, results in an increase in the excretion of bound amino acids, particularly hydroxyproline.[106] Individuals who eat shellfish may excrete an increased amount of taurine.

Aminoaciduria Due to Medication

The ingestion of certain drugs, particularly antibiotics, has been known to result in the excretion of ninhydrin-reactive compounds in the urine. Because of the increasing number of new drugs on the market, a careful drug history should be taken when an unusual amino acid pattern is found in the urine. Some of these compounds run to positions in close proximity to those amino acids involved in metabolic disorders and may, therefore, cause confusion. As a general rule, the authentic amino acid should be chromatographed together with the urine to show their dissimilarity. Drug ingestion as the cause of the abnormal aminoaciduria should be ruled out before embarking on extensive investigation for a possible new "disorder." It is advisable that a study of the urine be repeated after the patient has been off all medication for at least three or four days. Individual drugs causing abnormal spots will be discussed.

Ampicillin causes several purple and brown spots when the chromatogram is treated with ninhydrin. In one-dimensional chromatography by BuAc these are in the region of the leucines, phenylalanine, and valine and may be confused with these amino acids. In the two-dimensional HVE-BuAc system these spots are ill-defined and streaky and are to the right of phenylalanine (Figure 2-29). The urine occasionally gives a weakly positive cyanide-nitroprusside reaction which is probably due to a conjugate of ampicillin. In patients given penicillin, penicillamine sulfonic acid can be detected after treatment of the urine with hydrogen peroxide.[107]

Kanamycin*** is an antibiotic often used in combination with ampicillin. It is excreted as a ninhydrin-reactive compound which migrates faster than lysine in the HVE and remains near the origin in BuAc (Figure 2-29).

Carbenecillin**** causes an ill-defined purplish area at the right upper corner in addition to one purplish spot to the right of phenylalanine and one to the right of glutamine on HVE-BuAc chromatogram (Figure 2-30).

Cephalexin***** is a relatively new antibiotic which gives a ninhydrin-purple spot. On a one-dimensional chromatogram developed in BuAc, it is almost indistinguishable from phenylalanine. In the two-dimensional electrochromatogram, cephalexin moves next to phenylalanine (Figure 2-31). A negative ferric chloride test does not always rule out the possibility of phenylalanine because patients with a hyperphenylalaninemia of 12 to 15 mg/100 ml or less excrete a detectable quantity of phenylalanine and still give a negative ferric chloride test. That these are two different compounds can be demonstrated by co-chromatographing the urine containing cephalexin with authentic phenylalanine. A history of drug ingestion and a measurement of blood phenylalanine will give an unequivocal answer.

*Prosobee®, Mead Johnson; Isomil®, Ross Laboratories; Neomullsoy®, Borden; and Cho-Free®, Syntex
**Pedameth®, Durst
***Kantrex®, Bristol Laboratories, Syracuse, N.Y.
****Geopen®, Roering Div., N.Y., N.Y.
*****Keflex®, Lilly Co.

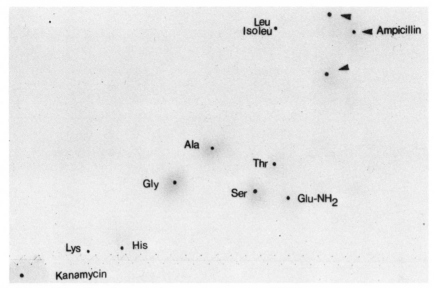

FIGURE 2-29. Two-dimensional paper chromatogram (HVE-BuAc) of urine from a patient receiving ampicillin and kanamycin.

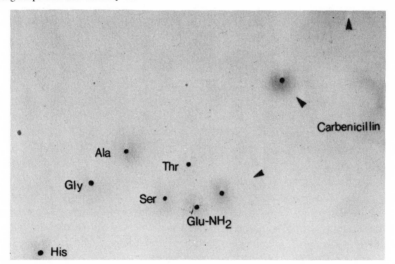

FIGURE 2-30. Two-dimensional paper chromatogram (HVE-BuAc) of urine from a patient receiving carbenicillin.

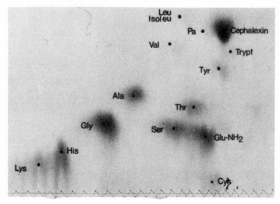

FIGURE 2-31. Two-dimensional paper chromatogram (HVE-BuAc) of urine from a patient receiving cephalexin.

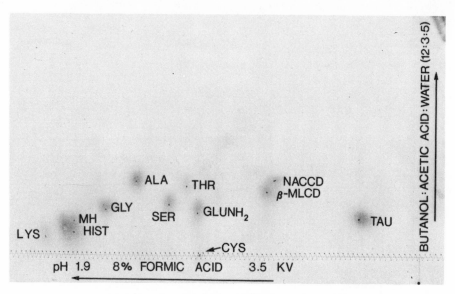

FIGURE 2-32. Two-dimensional paper chromatogram (HVE-BuAc) of urine showing the excretion of *N*-acetylcystine by a patient given *N*-acetylcysteine. *β*-Mercaptolactate-cysteine disulfide was added to the urine to illustrate the similar mobility of these two compounds in this system. (Reproduced from Shih, V. E. and Schulman, J. D., *N*-Acetylcysteine-cysteine disulfide excretion in the urine following *N*-acetylcysteine administration, *J. Pediatr.*, 74, 129, 1969, with permission from the author and the publisher.)

One patient who received isoniazid therapy had *β*-alanine in the urine.[108]

Abnormal amino acid excretion has been reported in patients with generalized psoriasis treated with high doses of 6-azauridine triacetate.[109] Urinary excretion of the *β*-amino acids (*β*-alanine, BAIB, taurine), *γ*-aminobutyric acid, and the sulfur amino acids (homocystine and homocysteine-cysteine disulfide) was increased. These abnormalities should be differentiated from the genetic disorders involving these amino acids.

Leukemic patients receiving folate antimetabolite* therapy and patients with vitamin B_{12} or folic acid deficiency may excrete 4(5)-amino-5(4)-imidazole carboxamide (AIC).[110] AIC has a mobility similar to *β*-aminoisobutyric acid, *β*-alanine, and *δ*-aminolevulinic acid, and reacts with ninhydrin to form a yellowish-brown color.

N-acetyl-cysteine** is a mucolytic agent administered by aerosol. This compound is absorbed through the mucosa and is excreted both in its oxidized form (*n*-acetyl-cystine) and as a mixed disulfide of *n*-acetyl-cysteine and cysteine.[111] Both derivatives react with the cyanide-nitroprusside reagent. The latter compound is also ninhydrin-reactive and moves to a location close to another cyanide-nitroprusside-positive compound, *β*-mercaptolactate-cysteine disulfide (BMLCD) (Figure 2-32), which is found in patients with a metabolic defect in sulfur amino acids (Table 1-1).

Diaper rash is a common problem in pediatrics and there are many commercial preparations available for its treatment; two of these contain amino acids. The oral medication with DL-methionine which causes D-methioninuria has already been discussed. Another preparation is a mixture of amino acids. Application of this cream before or during the collection of filter paper urine specimen causes distortion of the amino acid pattern in the urine specimen. The leucine and methionine spots are the most prominent (Figure 2-33). On a one-dimensional chromatogram developed in BuAc this pattern resembles that seen in maple syrup urine disease (MSUD) as valine and methionine overlap. Methionine can be identified by the following methods: (1) the characteristic odor of methionine resembles beef boullion and is outstanding on a freshly prepared chromatogram

*Methotrexate®, Lederle
**Mucomyst®, Mead Johnson

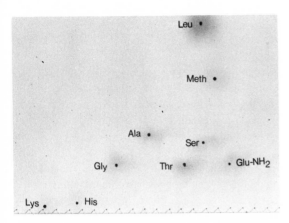

FIGURE 2-33. Two-dimensional paper chromatogram (HVE-BuAc) of urine showing greatly increased leucine-isoleucine spot and methionine spot as a result of contamination from diaper cream (Methakote®).

and (2) on a two-dimensional chromatogram, valine and methionine appear as two separate spots. Another point of differentiation between MSUD and Methakote* artifact on a one-dimensional chromatogram is that the leucine-isoleucine spot is always more intense than the lower valine-methionine spot in MSUD, while these two spots are usually of equal intensity in the Methakote artifact. Obviously, the blood amino acid pattern is normal in Methakote artifact.

The increasing use of L-arginine in clinical medicine, either as a diagnostic agent, e.g., growth hormone stimulation test, or as a therapeutic agent, e.g., in the treatment of cystic fibrosis, may cause some confusion in the interpretation of the results of urinary amino acid study. Administration of large doses of L-arginine results in an elevation of the blood arginine level and marked urinary excretion of arginine, ornithine, lysine, and cystine. Increased excretion of the three additional amino acids which share the same renal transport system with arginine is due to competitive inhibition of the renal tubular reabsorption of these amino acids by arginine. This pattern resembles that seen in homozygous cystinuria, hyperlysinemia, and hyperargininemia. Again, a history, measurement of blood amino acids and ammonia, and a repeat study are important in making the differentiation.

Intravenous infusion of an amino acid mixture

(hyperalimentation) to sustain the life of a patient has become an accepted practice in clinical medicine. Rapid infusion or overdose may result in a generalized hyperaminoaciduria, or if certain amino acids are given in excess, they may appear in the urine in increased quantity. If the DL form of amino acids is used, most of the D-amino acids will appear in the urine.

Aminoaciduria Due to Contamination

The most common cause of contamination of filter paper urine specimens of infants is feces. The contaminated filter paper often appears brownish, and specks of feces may be seen. Feces contain considerable amounts of free amino acids. Mixing of stool and urine causes a generalized amino-aciduria pattern with large amounts of proline and glutamic acid[112] (Figure 2-34). This artifact may mask or obscure real abnormalities in the urine.

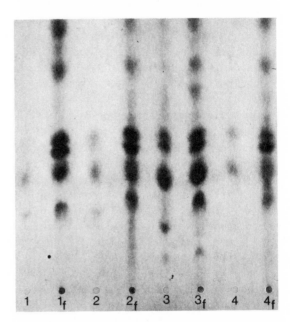

FIGURE 2-34. Amino acid chromatogram showing the effect of fecal contamination. 1, 2, 3, and 4 represent clean, uncontaminated filter paper urine specimens. 1f, 2f, 3f, and 4f represent the corresponding filter paper specimens contaminated with feces. There is a much greater concentration of free amino acids present in the contaminated specimens than in the uncontaminated ones. (Reproduced from Levy, H. L. et al., Fecal contamination in urine amino acid screening: Artifactual cause of hyperaminoaciduria, *Am. J. Clin. Pathol.,* 51, 765, 1969, with permission from the author and publisher.)

*Methakote®, Borden

Therefore, another specimen should be requested and the screening repeated.

There is another source of contamination worth mentioning. At the author's laboratory we have often noticed and been puzzled by the sudden appearance of "showers" of tiny ninhydrin-reactive spots all over the chromatogram. Recently, we realized that the appearance of the artifact coincides with the preparation of a synthetic amino acid diet. Apparently when large quantities of amino acids are scooped out of the bottle and weighed on a scale in the same room where chromatograms are prepared, particles of these amino acids settle on the filter paper and result in ninhydrin-reactive "spots." Amino acids should be weighed in a separate room, and technicians should change their laboratory coats before preparing chromatograms.

Improper storage or bacterial contamination of the specimen may result in the disappearance of amino acids, principally, serine and threonine. Prolonged storage of the specimen even at temperatures below 0°C may result in the decomposition of glutamine to glutamic acid, and asparagine to aspartic acid.

Increased Phenylalanine Concentration

Phenylalanine is not detectable on normal urine chromatograms; a definite spot indicates elevation. Increased phenylalanine concentrations in the blood and urine signify a metabolic block in the conversion of phenylalanine to tyrosine, or of tyrosine to other metabolites. In the former, tyrosine concentration is either low or normal, whereas in the latter the elevation of tyrosine is usually greater than that of phenylalanine.

When elevated phenylalanine and normal tyrosine levels (Figure 2-5) are detected in a newborn blood specimen, it should be reported immediately and a plasma or serum specimen should be obtained for quantitative measurement in order to confirm or rule out phenylketonuria (PKU), a block in the conversion of phenylalanine to tyrosine (Table 1-1). A rapid rise of blood phenylalanine concentration to greater than 20 mg/100 ml in the first week of life is indicative of PKU. Once the diagnosis has been made, the infant should be treated as early as possible. The urine ferric chloride test is unreliable for detection of PKU in early infancy because of the delayed maturation of the transaminase. In infants with severe atypical PKU or hyperphenylalaninemia,

blood phenylalanine levels reach 20 mg/100 ml at a later age (2 or 3 weeks of age or older), or they may never go as high as 20 mg/100 ml. The differential diagnosis of typical or atypical PKU may not be possible for quite a few months. In the so-called persistent mild hyperphenylalaninemia (PMH), phenylalanine levels are usually less than 12 mg/100 ml. These patients have a negative ferric chloride test at all times, but a phenylalanine spot is detectable on the urine amino acid chromatogram (Figure 2-35).

A semiquantitative estimation of the phenylalanine level can be performed on one-dimensional chromatograms. In BuAc, no other amino acid overlaps phenylalanine. A 3/16 in. disc and a 1/4 in. disc of the filter paper specimen, or 5 and 10 μl plasma are applied to the paper, and graded standards of phenylalanine 2.5, 5, 7.5, 10, 15, 20 mg/100 ml are spotted on the same paper.

As discussed in the previous section, the cephalexin spot should not be mistaken for phenylalanine.

Increased Concentrations of Tyrosine and Its Metabolites

A tyrosine elevation can easily be detected in both blood and urine (Figures 2-5 and 2-36). Tyrosine metabolites, p-hydroxyphenylacetic acid (PHPAA) and p-hydroxyphenyllactic acid (PHPLA), move with the solvent front in BuAc and are located at the right upper corner on the electrochromatogram. These metabolites can be

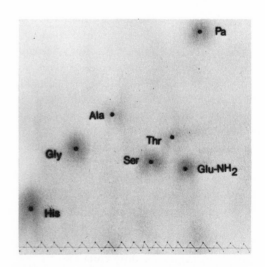

FIGURE 2-35. Two-dimensional paper chromatogram (HVE-BuAc) of urine from a patient with phenylketonuria showing a spot of phenylalanine.

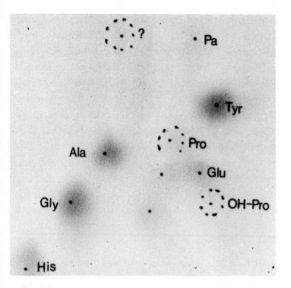

FIGURE 2-36. Two-dimensional paper chromatogram (HVE-BuAc) showing the urinary amino acid pattern in transient neonatal hypertyrosinemia. Note the presence of increased amounts of tyrosine and phenylalanine, and an unknown ninhydrin-yellow spot above alanine.

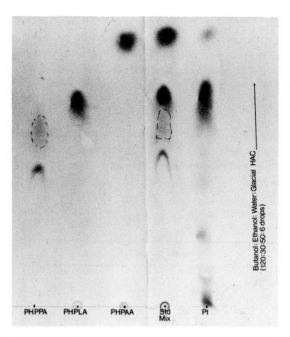

FIGURE 2-37. One-dimensional chromatogram of tyrosine metabolites developed in butanol:ethanol:water: glacial acetic acid (120:30:50:6 drops) and stained with diazotized sulfanilic acid reagent (Pauly reagent). Abbreviations: PHPPA, *p*-hydroxyphenylpyruvic acid (orange); PHPLA, *p*-hydroxyphenyllactic acid (purple); PHPAA, *p*-hydroxyphenylacetic acid (purple); Std Mix, standard mixture; Pt., urine from a patient with hereditary tyrosinemia.

better separated in other solvents, such as butanol-pyridine-water (14:3:3), benzene-acetic acid-water (125:75:3), and butanol-ethanol-water-acetic acid (120:30:50:6 drops) (Figure 2-37), and can be visualized by dipping the chromatogram into diazotized sulfanilic acid reagent; spots first appear reddish in color and then turn to a permanent purple color. The nitrosonaphthol test gives additional evidence of the presence of *p*-hydroxyphenolic acids.

In a newborn infant, hypertyrosinemia and tyrosyluria are most likely the benign type of transient neonatal hypertyrosinemia. When these are found in older infants or children they are more likely pathologic. In the hereditary tyrosinemia associated with liver disease and renal tubular dysfunction, blood tyrosine elevation is usually in the intermediate range (about 10 mg/100 ml or 555 μmol/l.), and the urine may show a generalized increase in amino acids with prominent tyrosyluria and reducing substances. At a later stage, blood methionine may also be increased. In the type of hereditary tyrosinemia without liver and renal involvement, the blood tyrosine concentration is markedly elevated to more than 15 mg/100 ml or 828 μmol/l. In the case of acquired hypertyrosinemia secondary to liver disease such as hepatitis, the elevation in blood tyrosine concentration may vary from mild to marked, and increased methionine concentration is frequently found.

It is difficult to distinguish the various causes of hypertyrosinuria and tyrosyluria on the basis of these laboratory findings. Differential diagnosis is based upon other findings.

Increased excretion of *p*-hydroxyphenylacetic acid by patients with cystic fibrosis has been reported.[113]

Homogentisic acid is a derivative of tyrosine and is excreted in large quantity by alkaptonuric patients. The urine may turn black on standing at alkaline pH. Homogentisic acid appears as a dark brown spot when the chromatogram for creatinine determination is stained with Jaffe reagent. This could be conveniently used as a screening test for alkaptonuria.

Tryptophan and Indole Abnormalities

Tryptophan stains poorly with ninhydrin. With a small quantity it may appear as a brownish-purple spot. Ehrlich reagent is much more sensitive for tryptophan. Indoles can best be studied by the two-dimensional paper chromatography using solvent combinations of IPrAm/BuAc or BuAc/KCl, as described earlier, and stained by Ehrlich reagent (Figures 2-38 A and B).

Normal urine contains very little tryptophan or indole. Armstrong et al.,[117] however, reported finding 38 indole compounds in urine. In Hartnup disease, urinary tryptophan as well as other neutral amino acids is increased; the amounts of urinary indoles and indican vary from normal to greatly increased. An isolated increase in urinary tryptophan without indoles has been found in the disorder tryptophanuria with dwarfism. In "blue diaper" syndrome, urinary indoles (indoleacetamide, indolelactic acid, indoleacetic acid, indoleacetylglutamine, and indican) are increased. The blue discoloration of the diaper is due to indigotin (indigo blue). It may also be caused by pigments produced by *P. aeruginosa* present in stools, or by contact of ammonial urine with copper diaper snaps.

Increased excretion of 3-hydroxykynurenine and kynurenine is found in hydroxykynureninuria (Figure 2-39) and in pyridoxine deficiency. In the latter, xanthurenic acid in the urine is also

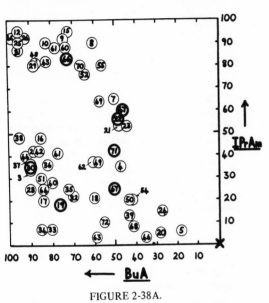

FIGURE 2-38A.

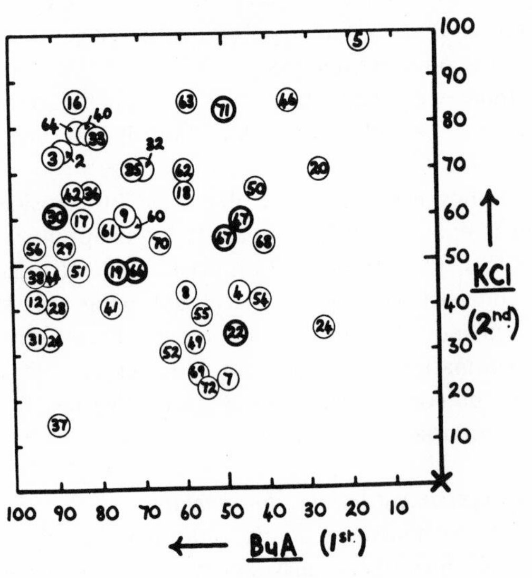

FIGURE 2-38B.

FIGURE 2-38. Map of indoles and related Ehrlich-reactors. (A) Chromatogram developed in IPrAm (isopropanol-ammonia-water, 200:10:20) followed by BuA (butanol-acetic acid-water, 12:3:5). (B) Chromatogram developed in BuA, followed by KCl (20% w/v). (2) *N*-acetyl-tryptophan; (3) Anthranilic acid; (4) Bufothionine; (5) Citrulline; (7) 5:6 Dimethoxy-tryptamine (as HCl); (8) *N:N*-Dimethyl-5-hydroxy-tryptamine (as acetate); (9) *N:N*-Dimethyl-tryptamine (as acetate); (10) $a:a$-Dimethyl-tryptamine (as acetate); (12) Ethyl indolyl-acetate; (15) Gramine; (16) Hippuric acid; (17) 3-Hydroxy-anthranilic acid; (18) 5-Hydroxy-anthranilic acid; (19) 5-Hydroxy-indolyl-acetic acid; (20) 5-Hydroxy-indolyl-acetic acid 5-sulfate ester; (21) 4-Hydroxy-tryptamine (as creatinine sulfate); (22) 5-Hydroxy-tryptamine (as creatinine sulfate); (23) 6-Hydroxy-tryptamine (as creatinine sulfate); (24) 5-Hydroxy-tryptophan; (25) Indole; (26) Indole-3-aldehyde; (27) Indole-2-carboxylic acid; (28) Indole-3-carboxylic acid; (29) Indolyl-acetamide; (30) 3-Indolyl-acetic acid; (31) 3-Indolyl-acetonitrile; (32) Indolyl-acetyl-asparagine; (33) Indolylacetyl-aspartic acid; (34) Indolylacetyl-glutamic acid; (35) Indolylacetyl-glutamine; (36) Indolylacetyl-glycine; (37) 3-Indolyl-acrylic acid; (38) a-(3-Indolyl)-butyric acid; (39) C-(3-indolyl) glycine; (40) 3-Indolyl-glycollic acid; (41) 3-Indolyl-glyoxylic acid; (42) 3-Indolyl-lactic acid; (43) Indolyl-propane 1:2 diol; (44) β-(3-indolyl)-propionic acid; (46) Indoxyl-0-phosphate; (48) Isatin; (49) Kynurenic acid; (50) Kynurenine; (51) 5-Methoxy-indolyl acetic acid; (52) 5-Methoxy-tryptamine (as HCl); (54) 5-Methoxy-tryptophan; (55) *N*-methyl-5-hydroxy-tryptamine (as oxalate); (60) *N*-methyl-tryptamine (as HCl); (61) a-Methyl tryptamine (as acetate); (62) *N*-methyl-tryptophan; (63) Porphobilinogen; (64) Pyrrole-2-carboxylic acid; (65) Skatole; (66) Tryptamine; (67) Tryptophan; (68) Tryptophan-4-carboxylic acid; (69) Tryptophanamide; (70) Tryptophanol; (71) Urea; and (72) Xanthurenic acid. (Reproduced from Jepson, J. B., Indoles and related Ehrlich reactors, in *Chromatographic and Electrophoretic Techniques, Volume I, Chromatography*, 3rd ed., Smith, I., Ed., John Wiley & Sons, New York, 1969, 243, with permission from the author and publisher.)

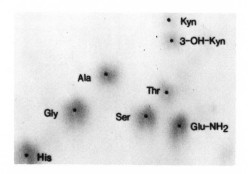

FIGURE 2-39. Two-dimensional paper chromatogram (HVE-BuAc) of urine showing the presence of 3-hydroxy-kynurenine (brown) and kynurenine.

increased. We have seen a transient presence of kynurenine and hydroxykynurenine in the urine of apparently healthy infants without a definite history of vitamin deficiency.

A large quantity of 5-hydroxyindoleacetic acid (5-HIAA), a metabolite of serotonin, is present in the urine of patients with carcinoid. Abnormal amounts of tryptophan metabolites have been reported in various neuropsychiatric disorders; however, no definite metabolic defect has been identified. The exogenous origin of these Ehrlich-reactive compounds[105] should always be ruled out before any significance can be attached to such findings.

Abnormalities in Histidine and Imidazole Derivatives

Histidine, methylhistidines, carnosine, and the basic amino acids migrate to the same location on a one-dimensional chromatogram. They appear as two spots in butanol-acetic acid with a filter paper blood or urine specimen. When serum is applied, only one spot is present. Histidine and carnosine can be differentiated from the basic amino acids by staining the chromatogram with diazotized sulfanilic acid reagent (Pauly reagent). These two compounds turn red in color, whereas the

methylhistidines and the basic amino acids do not react.

In histidinemia the blood histidine level is usually five to ten times normal. The elevation can be detected in newborn blood by the bacterial inhibition test.[118] In addition to a large histidine spot, the urine of a histidinemic person contains two imidazole metabolites of histidine, imidazole-lactic acid (ILA), and imidazoleacetic acid (IAA) which can be visualized with Pauly reagent. ILA is at the level of alanine, while IAA is just below tyrosine which also reacts with Pauly reagent. It is not always possible to recognize the specific histidine metabolites on a one-dimensional chromatogram since other unidentified Pauly-reactive imidazole compounds have similar R_f values and run into the same locations. However, the pattern of histidinemia on a two-dimensional chromatogram is characteristic. When a ninhydrin-treated chromatogram shows any increase in the size and intensity of the spot containing histidine and methyl-histidines (Figure 2-40), it should be overstained with Pauly reagent. Young infants with histidinemia may excrete large amounts of histidine but little of the metabolites. Figure 2-41 illustrates the locations of histidine metabolites and known and unknown Pauly-reactive ninhydrin-negative compounds that have been found in the urine.

Histidase activity is present in normal skin and has been found to be deficient in patients with histidinemia. A fairly large quantity of skin is required for measurement of the enzyme activity; it is almost impossible to get this amount from young children. Histidase deficiency can easily be demonstrated indirectly by a simple chromatographic method[119] using less than 0.5 mg of skin.*

Carnosine and anserine are inseparable from lysine in the HVE at pH 1.6, and both stain brown or sometimes purple on the two-dimensional HVE-BuAc chromatogram. Carnosine reacts with Pauly reagent, while anserine does not. Anserine is an exogenous amino acid; the most common

---

*Small pieces of cuticle are obtained with a nail clipper or fine scissors from the areas of thickened skin, either around fingernails or toenails. Sweat contains large amounts of urocanic acid. Therefore, the areas should be wiped clean before manipulation, and the specimens should be picked up with forceps to avoid contamination from the hand of the physician. The pieces of skin are placed in a mortar and homogenized with a pestle in several drops of concentrated ammonium hydroxide. As it evaporates, more ammonium hydroxide may be added until the skin is properly ground. The homogenate is spotted on a piece of Whatman 3MM paper with histidine and urocanic acid standards. It is chromatographed in BuAc overnight and stained with diazotized sulfanilic reagent. The skin from normal persons contains abundant amounts of urocanic acid and histidine, whereas that from histidinemic patients contains large amounts of histidine but no urocanic acid (Figure 2-42). As little as 0.2 μg urocanic acid can be visualized.

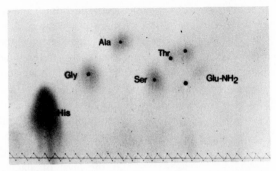

FIGURE 2-40. Two-dimensional paper chromatogram (HVE-BuAc) of urine from a patient with histidinemia showing markedly increased histidine.

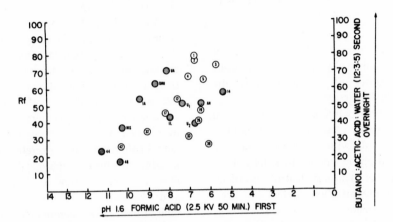

FIGURE 2-41. Map of imidazole derivatives which may be found in urine. Abbreviations: AH: *N*-acetyl-histidine; DHU: dihydrourocanic acid; HIS: histidine; IA: imidazoleacetic acid; IL: imidazolelactic acid; UA: urocanic acid; U₁: unknown No. 1; U₂: unknown No. 2. All these compounds (shaded circles) appear as red spots with diazotized sulfanilic acid (Pauly reagent). For reference, amino acids are shown as numbered circles, which correspond to those in Figure 2-28.

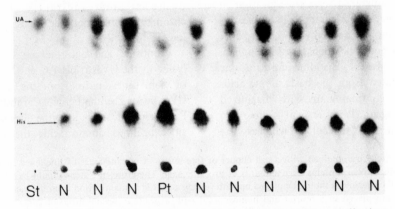

FIGURE 2-42. Demonstration of the absence of urocanic acid as an indication of histidase deficiency. One-dimensional paper chromatogram of skin homogenate developed in BuAc and stained with diazotized sulfanilic acid reagent. UA: urocanic acid; HIS: histidine; N: normal person; PT: patient with histidinemia; ST: urocanic acid standard.

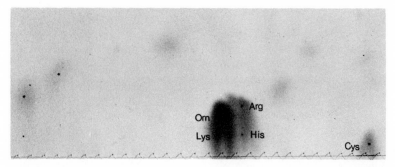

FIGURE 2-43. Two-dimensional paper chromatogram (HVE-BuAc) of urine showing the markedly increased amounts of cystine, lysine, ornithine, and arginine in homozygous cystinuria. Also note two fast-moving amines (far left) often seen in this disorder.

dietary source is the white meat of poultry. Both compounds are increased following the ingestion of meat. Patients with the rare hereditary disorder, carnosinemia (carnosinase deficiency), excrete large amounts of carnosine even when on a meat-free diet. Carnosinase readily hydrolyzes carnosine to histidine and β-alanine, and anserine to 1-methylhistidine and β-alanine; thus, patients with carnosinase deficiency cannot hydrolyze anserine as normal individuals do. The absence of 1-methylhistidine in urine containing anserine is a clue to the diagnosis. A definitive diagnosis can be obtained by assaying carnosinase activity in serum.[120,121]

Abnormalities in Sulfur Amino Acids

Sulfur amino acids, except for taurine and a small amount of cystine, are usually not found in normal urine. A positive cyanide-nitroprusside test is indicative of an abnormality. Cystine is the compound most frequently responsible for the positive test. On a one-dimensional chromatogram, it runs to the same position as the basic amino acids. On a two-dimensional chromatogram, it is a distinct spot.

Cystine and the basic amino acids (lysine, arginine, and ornithine) share a common renal transport mechanism. All four amino acids are increased in the urine of a homozygote for the renal transport defect, cystinuria (Figure 2-43). In a heterozygote for this trait only lysine and cystine are increased (Figure 2-44). Occasionally, it is difficult to differentiate between these two conditions by paper chromatography; a quantitative measurement of the excretion of these compounds on an amino acid analyzer is necessary. Homozygous cystinuric patients on

penicillamine (dimethylcysteine) therapy show two additional ninhydrin-purple spots, both above cystine and slightly toward the origin on a two-dimensional electrochromatogram. One of these is the disulfide of penicillamine itself, and the other a mixed disulfide of penicillamine and cysteine (Figure 2-45).

Patients with hyperlysinemia excrete increased amounts of cystine as well as arginine and ornithine due to competitive inhibition of the transport system (Figure 2-46). For the same reason, the urine of patients with hyper-argininemia and of those receiving large doses of arginine has the same amino acid pattern. It would appear that cystinurics excrete proportionately more cystine and that hyperlysinemics excrete more lysine, while hyperargininemics excrete more arginine. In order to make the correct interpretation, a history of medication, blood screening, and quantitative measurement of amino acid levels in blood and urine should be considered along with the urine findings.

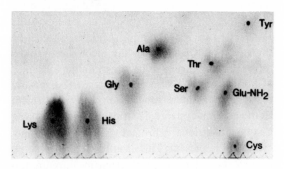

FIGURE 2-44. Two-dimensional paper chromatogram (HVE-BuAc) showing the urinary amino acid pattern in heterozygous cystinuria. Note the increased amount of lysine and cystine without the presence of arginine and ornithine.

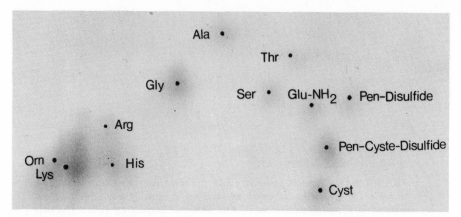

FIGURE 2-45. Two-dimensional paper chromatogram (HVE-BuAc) of urine from a patient with cystinuria on penicillamine therapy, showing the excretion of penicillamine disulfide, penicillamine-cysteine mixed disulfide, as well as cystine and the basic amino acids.

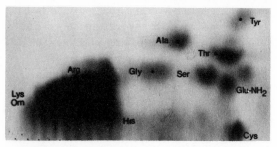

FIGURE 2-46. Two-dimensional paper chromatogram (HVE-BuAc) showing the urinary amino acid pattern in hyperlysinemia. Note the presence of arginine and ornithine and cystine in addition to lysine.

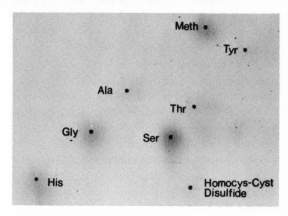

FIGURE 2-47. Two-dimensional paper chromatogram (HVE-BuAc) of urine from a patient with cystathionine synthase deficiency, showing the intense serine-homocystine spot and the presence of methionine and homocysteine-cysteine disulfide.

Homocystine moves to the same position as serine and can be easily missed without the aid of the nitroprusside test. In the absence of cystine, a positive nitroprusside test and a large "serine spot" on the two-dimensional chromatogram (Figure 2-47) are highly suggestive of the diagnosis of homocystinuria. A one-dimensional chromatogram, with homocystine and cystine standards run in parallel as markers, should be prepared and stained with the alcoholic cyanide-nitroprusside reagent to demonstrate the existence of homocystine. A silver-nitroprusside test specific for homocystine can also be performed. Homocysteine-cysteine disulfide is usually seen on a two-dimensional chromatogram when homocystinuria is present. This disulfide is also excreted by cystinuric patients, but often in amounts too low to be detected on paper.

Due to the high renal clearance, homocystine accumulation in blood is too low to be detected by paper chromatography. On the other hand, if

methionine is elevated, it is usually much greater in the blood than in the urine. In cystathionine synthase deficiency, hypermethioninemia is a prominent feature in early infancy (Figure 2-30), and homocystinuria may be of such a mild degree in the first month of life that it gives a negative or weakly positive cyanide-nitroprusside test. Thus, blood screening for methionine is much better than urine screening for abnormal sulfur amino acids to detect this metabolic defect in newborns. As the patient grows older, hypermethioninemia decreases and homocystinuria increases. In some adult patients methionine elevation can only be detected by quantitative measurement. A normal or below normal blood methionine concentration in a patient who excretes both homocystine and cystathionine suggests other types of homo-

TABLE 2-7

Differential Diagnosis of Genetic Homocystinurias by Laboratory Findings

	Blood		Urine		
Enzyme defect	Methionine	Cystathionine	Cystathionine	Methyl-malonic acid	Inorganic sulfate excretion after methionine load
Cystathionine synthase	Increased	Absent	Absent	Absent	Below normal
B_{12}-coenzyme metabolism	Normal or low	May be present	Present	Present	Normal
$N^{5,10}$-Methylenetetra-hydrofolate reductase	Normal or low	May be present	Present	Absent	Normal

cystinuria, namely defects in the methylation of homocysteine (Table 2-7). An additional finding of methylmalonic acid in the urine distinguishes B_{12} coenzyme metabolic defect from the methylene-tetrahydrofolate reductase deficiency.

With one-dimensional chromatography it is not easy to distinguish cystathionine from other amino acids by ninhydrin stain. Platinic iodide stain can be useful as a preliminary test. Cystathionine appears as a distinct spot on a two-dimensional HVE-BuAc chromatogram (Figure 2-48). Blood screening is ineffective in the detection of this compound since only a small amount is accumulated in the blood.

Cystathionase deficiency is a rare benign metabolic disorder (Table 1-1) in which cystathionine is excreted in the urine in large quantity. Cystathioninuria is occasionally found in normal infants and may also be present in other disease states. It is excreted by patients with functional neural tumors, e.g., neuroblastoma and ganglioneuroblastoma[121,122] and hepato-blastoma,[124] and this finding can be used as a diagnostic aid. Only about half of the patients with neuroblastoma have cystathioninuria; its presence is independent of the increased vanil-mandelic acid (VMA) excretion.[123] Thus, the absence of cystathioninuria does not rule out these tumors. Since these neoplasms should be treated as soon as possible, there should be no delay in reporting results. Cystathioninuria has also been reported in patients with liver disease.[125]

β-mercaptolactate-cysteine disulfide gives a positive nitroprusside test; it has an R_f similar to

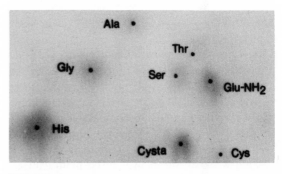

FIGURE 2-48. Two-dimensional paper chromatogram (HVE-BuAc) of urine from a patient with cystathioninuria.

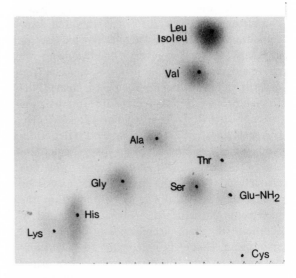

FIGURE 2-48A. Two-dimensional paper chromatogram showing the increased amounts of leucine, isoleucine, and valine in the urine of an infant with maple syrup urine disease.

that of homocystine in BuAc and can be easily confused with acetylcysteine-cysteine disulfide in the HVE and BuAc systems (Figure 2-32). When in doubt, the urine should be chromatographed with authentic acetylcysteine-cysteine disulfide. Alternatively, a duplicate two-dimensional chromatogram should be made and stained with the alcoholic cyanide-nitroprusside reagent; the appearance of ninhydrin-negative N-acetylcystine as a purple-red spot at the upper right corner of the chromatogram indicates that the abnormal spot is secondary to the medication, N-acetylcysteine.

S-Sulfocysteine is excreted in sulfite oxidase deficiency (Table 1-1) and is located in the vicinity of taurine and cysteic acid. The presence of an increased amount of sulfite and thiosulfate in the urine detectable by a spot test supports the diagnosis.

Abnormalities in Branched Chain Amino Acids

Leucine and isoleucine overlap and cannot be separated in either BuAc or HVE. Both these amino acids and valine are absent in normal urine. Their presence usually indicates pathology.

MSUD is a medical emergency in the neonatal period because untreated infants may die in the second week of life or may suffer brain damage. Marked accumulation of these branched chain amino acids in blood is evident by the fifth day of life when blood for routine screening is usually taken and can be easily detected by paper chromatography or by bacterial inhibition assay for leucine. At this age, urine screening may not be as reliable as blood screening. On one-dimensional chromatography, both the leucine-isoleucine spot and the valine-methionine spot are increased, the former being larger and more intense than the latter (Figure 2-5). On a two-dimensional electro-chromatogram, valine is a spot distinct from methionine. The presence of the maple syrup odor and a positive DNPH test in the urine are characteristic of MSUD. Amino acid analysis by ion-exchange chromatography will confirm the increases in the branched chain amino acids and, in addition, will show the presence of an abnormal amino acid, alloisoleucine (Figure 2-17) which is not separated from leucine and isoleucine by paper chromatography. The ketoacids can be studied by paper chromatographic or TLC methods described in Chapter 4. It should be cautioned that the amino acid abnormalities in the intermittent type

of MSUD are only seen during the attacks, and a normal amino acid pattern while the patient is asymptomatic does not negate the diagnosis. If history suggests the diagnosis, repeat amino acid studies should be performed during an attack, and the enzyme should be measured in leukocytes or cultured fibroblasts. Recently, other variants of MSUD have been reported; in one of these, nonspecific mental retardation without keto-acidosis was the only finding. Thus, the absence of ketoacidosis does not rule out MSUD.

The accumulation of valine alone in blood and urine strongly suggests valine transaminase deficiency (hypervalinemia) (Table 1-1) in which the ketoacid of valine is not increased.

In other defects in branched chain amino acid metabolism, abnormal organic acids but no amino acid abnormalities are found; these disorders will be discussed in Chapter 4.

It should be noted that mild elevations (two to three times normal) of the branched chain amino acids in blood have been observed in a normal person after three or four days' starvation.[126]

Abnormalities in Urea Cycle Intermediates and Glutamine

The four urea cycle intermediates, ornithine, citrulline, ASA, and arginine, can be identified on the paper chromatogram by location and staining characteristics. All but ASA are present in normal blood in small quantities. However, none of the four compounds is excreted in the normal urine in amounts detectable by chromatographic screening methods.

In the two disorders with blocks of carbamyl-phosphate synthetase and ornithine carbamyl-transferase (the congenital hyperammonemias), blood ammonia is markedly increased. However, the only amino acid abnormality is an increased glutamine concentration. Glutamine elevation is a nonspecific finding and may be seen in any condition in which hyperammonemia is present. Thus, the diagnosis is suggested by the clinical history and the finding of marked hyper-ammonemia is confirmed by the liver enzyme assay.

In citrullinemia (Figure 2-49), citrulline is greatly increased in both blood and urine and is easily detectable by paper chromatography. It is mildly increased in argininosuccinic aciduria. Citrulline stains yellow with Ehrlich reagent, but it appears as a peach-colored spot when a ninhydrin-

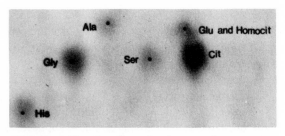

FIGURE 2-49. Two-dimensional paper chromatogram (HVE-BuAc) showing increased excretion of citrulline in the urine of a patient with citrullinemia.

treated chromatogram is dipped in Ehrlich reagent.

In argininosuccinase deficiency, ASA is excreted in great quantity; most patients excrete over 2 g in 24 hr. Very little ASA is reabsorbed by the renal tubule, and the amount that is present in the blood may not be obvious on the one-dimensional paper chromatogram. However, the diagnosis can readily be made by urine screening. Free ASA is an unstable compound and it has a tendency to form cyclic anhydrides at acid pH. On the one-dimensional chromatogram developed in BuAc, these anhydrides move to the area just above arginine which is usually devoid of any ninhydrin-positive compounds (Figure 2-27). Other amino acids are generally present in less than normal quantities. By two-dimensional separation, ASA and its anhydrides appear as two or three partially separated spots (Figure 2-50). To the inexperienced eye, it looks almost like a contamination or an artifact. Once familiar with the pattern, the diagnosis cannot be mistaken. Confirmation of the diagnosis can be made by assaying argininosuccinase activity in the red blood cells, as described by Tomlinson and Westall,[127]

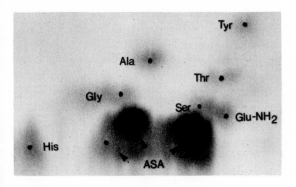

FIGURE 2-50. Two-dimensional paper chromatogram (HVE-BuAc) showing the urine amino acid pattern in a patient with argininosuccinic aciduria. Argininosuccinic acid and its anhydrides are excreted in excessive amounts.

or in cultured fibroblasts.[128] Since this disorder may present itself in the neonatal period and can cause early death, newborn screening by the simple bacterial auxotroph assay using filter paper blood is warranted. This enzyme assay can be performed on cord blood for early detection and early treatment.

Arginase deficiency results in arginine and ammonia accumulation in the blood. The amino acid abnormalities in the urine resemble those in homozygous cystinuria, namely increased amounts of arginine, as well as ornithine, lysine, and cystine, apparently as a result of competitive inhibition of the renal tubular reabsorption of these amino acids by arginine. As indicated previously, infusion or ingestion of arginine may result in alterations of blood and urine amino acid patterns indistinguishable from arginase deficiency.

Hyperornithinemia is one of the few disorders which cannot be diagnosed by examining the urine. In the type of hyperornithinemia without liver and renal involvement (Type 1), the blood ornithine concentration is greatly elevated (Table 1-1) and its presence as an unusually intense spot in the area of the basic amino acids can be detected on the one-dimensional chromatogram. Three other amino acids which move to the same area should be differentiated. Histidine and arginine can be ruled out by their reaction to Pauly reagent and Sakaguchi reagent, respectively. Analysis of the blood on an amino acid analyzer distinguishes lysine from ornithine. The urine contains very little ornithine. However, homocitrullinuria, which is of endogenous origin, is quite obvious on a two-dimensional chromatogram (Figure 2-51). Homocitrulline, with mobility similar to glutamic acid, can be identified by

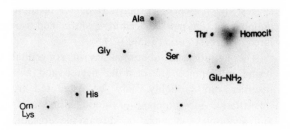

FIGURE 2-51. Two-dimensional paper chromatogram (HVE-BuAc) of urine from a patient with hyperornithinemia, hyperammonemia, and homocitrullinuria, showing the intense spot of homocitrulline and normal ornithine-lysine spot.

overstaining a ninhydrin- or isatin-treated chromatogram with Ehrlich reagent. As the chromatogram dries, a lasting pink spot of homocitrulline appears. Hyperammonemia and increased glutamine concentration are supportive evidence of this disorder.

The mild degree of elevation of blood ornithine in Type II hyperornithinemia associated with hepatorenal disease (ornithine-ketoacid transaminase deficiency) (Table 1-1) can only be found by quantitative measurement of blood amino acids on the analyzer. Such quantitative analysis of blood amino acids should therefore be considered in any infant with hepatitis of unknown etiology or of familial origin.

Markedly increased lysine and mildly increased arginine in the urine in the presence of below-normal concentrations of these amino acids in blood and hyperammonemia are suggestive of familial protein intolerance (FPI) (Table 1-1).

Abnormalities in Lysine and Its Metabolites

Blood lysine elevation, as discussed in the preceding paragraph, is indistinguishable from ornithine elevation on a paper chromatogram. Therefore, the two disorders, hyperornithinemia and hyperlysinemia, can only be distinguished by the urinary amino acid pattern. In the former, homocitrulline is found, while in the latter a pattern of increased lysine, arginine, ornithine, and cystine is present. The urinary amino acid pattern in hyperlysinemia (Figure 2-46) is similar to that in homozygous cystinuria (Figure 2-43). This again indicates that the so-called "cystinuria pattern" may be the expression of several different underlying disorders.

Increases in the excretion of the dibasic amino acids without a concomitant increase in cystine and vice versa with normal blood amino acids suggest the presence of transport defect of these amino acids, hyperdibasicaminoaciduria and isolated hypercystinuria (Table 1-2).

The amino acid abnormalities in congenital lysine intolerance are of mild degree and this disorder can probably only be diagnosed by quantitative measurement of blood amino acids and ammonia concentration during the episodic attacks.

In saccharopinuria, derivatives of lysine other than saccharopine, including homocitrulline,

homoarginine, and aminoadipic acid, are also found in the urine. Lysine and citrulline are increased several times the normal level in blood. Saccharopine has a low R_f in BuAc and moves to a position near cystine on a two-dimensional chromatogram (Figure 2-28).

Pipecolate is not separated from methionine on the one-dimensional chromatogram, but appears as a distinct spot on a two-dimensional chromatogram. With the regular ninhydrin reagent, it stains poorly giving a bluish-purplish color and a red fluorescence under ultraviolet light. With isatin reagent it is grayish-blue. The color yield of pipecolate can be intensified by a modification of the ninhydrin reagent.*[129] However, the disorder hyperpipecolatemia probably cannot be detected by screening because the amount of pipecolic acid present in blood and urine is too small to be detected by routine paper chromatography.

Hydroxylysine migrates to the same position as lysine in HVE and BuAc, and the amount excreted by patients with hydroxylysinuria is not easily detectable. Other solvent systems or ion-exchange chromatography will have to be used for diagnosis when the presence of hydroxylysine is suspected.

Increased Glycine and Sarcosine

Glycine is present in large quantity in normal urine, and the normal range is wide. Infants under six months of age usually excrete more than older children and adults. Therefore, it is even more difficult to decide whether an abnormal amount of glycine is excreted. In general, an increase in glycine associated with the presence of a moderate amount of proline, hydroxyproline, and other amino acids is most likely a normal "baby pattern." The true increased glycine in urine often appears as the only prominent spot on the chromatogram (Figure 2-52). Other amino acids may be normal or low.

Isolated hyperglycinuria with a normal blood glycine level usually is of little clinical significance. A heterozygote for the renal transport defect, iminoglycinuria, excretes increased amounts of glycine. It is occasionally related to the occurrence of rickets or muscle disease.

When increased glycine is found in the urine, it is important to screen the blood for hyperglycinemia. Increased glycine in the blood may be easily missed with paper chromatography; there-

*A solution of ninhydrin 0.2% (w/v) in glacial acetic acid-acetone (20:80).

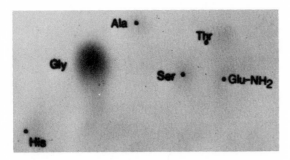

FIGURE 2-52. Two-dimensional paper chromatogram (HVE-BuAc) showing the markedly increased glycine in the urine of a patient with nonketotic hyperglycinemia.

fore, a quantitative measurement should be performed on an analyzer. In the nonketotic type of hyperglycinemia there is usually a moderate to marked increase in the blood glycine concentration, whereas in the ketotic type of hyperglycinemia, the blood glycine concentration is only mildly elevated. In the latter condition, hyperglycinemia is secondary; the primary defect may be a block in the metabolism of propionic acid or other fatty acids (Table 1-5). Chromatographic screening of the profile of short chain fatty acids or organic acids is advisable (see Chapter 4).

Sarcosine is not a normal constituent of urine. Its presence indicates the possibility of sarcosinemia. Blood amino acids may appear normal on paper chromatography because of the low concentration of sarcosine in the blood. This compound migrates to a position near alanine, being a little slower in both BuAc and HVE (Figure 2-53). The ninhydrin color of this spot is stable and remains visible when the chromatogram is left at room temperature for several days and after the color of all the other amino acids has faded.

Increased Proline, Hydroxyproline, and Glycine

Proline reacts with ninhydrin to form a yellow color which is barely visible with one-dimensional

paper chromatography of normal blood. Any obviously visible yellow color indicates proline elevation and requires further investigation. Other amino acids also stain with isatin and form different colors. However, proline is in an area devoid of other amino acids. Alanine, which stains blue and is just below proline in the one-dimensional chromatogram, may be confused with proline if it is present in a moderate to large quantity. A useful test for proline has been described by Pasieka and Morgan;[130] when the isatin-treated chromatogram is dipped in INHCl and washed with water, only the blue spot of proline persists.

Hydroxyproline stains poorly with both ninhydrin and isatin, and a yellow color spot may be seen on a ninhydrin-treated two-dimensional chromatogram when a large amount is present. It is best visualized when the chromatogram previously treated by either of the above-mentioned reagents is overstained with Ehrlich reagent. A red-purple spot develops in less than 1 min, and usually fades within a few hours.

Proline and hydroxyproline are not normally found in the urine of children over six months of age. When large amounts of these two imino acids, in addition to glycine, are found in the urine (Figure 2-54), one of three disorders is suggested: the renal transport defect, iminoglycinuria, or the disorders in proline metabolism, hyperprolinemia, Type I or Type II. Patients with iminoglycinuria excrete greater amounts of proline than normal infants and this disorder has been detected by urine screening of newborn infants four to six weeks of age. However, a definite diagnosis should be withheld until persistent iminoglycinuria has

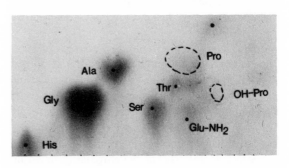

FIGURE 2-54. Two-dimensional paper chromatogram (HVE-BuAc) of urine showing the increased amounts of proline, hydroxyproline, and glycine, a pattern seen in both iminoglycinuria (renal transport defect) and hyperprolinemia.

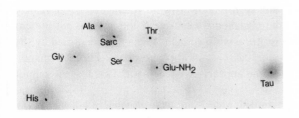

FIGURE 2-53. Two-dimensional paper chromatogram (HVE-BuAc) showing the presence of sarcosine in urine.

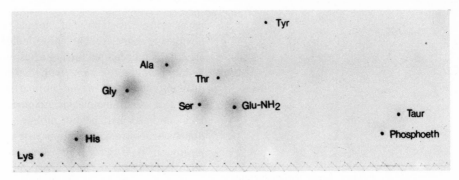

FIGURE 2-55. Two-dimensional paper chromatogram (HVE-BuAc) showing the presence of phosphoethanolamine in the urine of a patient with hypophosphatasia.

been demonstrated. Carriers of this trait may have hyperglycinuria. This transport defect is much more common than the hyperprolinemias.[131] In the latter conditions, blood proline is elevated. High concentrations of blood proline competitively inhibit the renal reabsorption of hydroxyproline and glycine sharing the same transport system. When the blood proline concentration reaches a level of 800 μmol/l. (9.2 mg/100 ml) or above, all three compounds are excreted in excessive amounts, a pattern resembling that seen in iminoglycinuria due to renal transport defect. Therefore, it is important to measure blood proline concentration to make the correct diagnosis. Type II hyperprolinemia can be differentiated from Type I by the excretion of Δ-pyrroline-5-carboxylic acid (PC). Its presence can be determined in fresh urine by its reaction with o-aminobenzaldehyde.*[132] The blood proline elevation is usually greater in Type II than in Type I hyperprolinemia. In those patients with moderate elevations, iminoglycinuria is not always present. Some of the carriers of Type I hyperprolinemia have mild elevation of blood proline which can only be detected by quantitative measurement.

Hydroxyproline without the presence of proline occasionally is found in the urine of normal infants between three and six months of age as the iminoglycinuria of infancy is disappearing. Its presence is certainly abnormal at any age after six months. Hydroxyproline has a high renal clearance and the diagnosis of hydroxyprolinemia by blood screening is not reliable. We chromatographed on

paper several different blood specimens from a patient with known hydroxyprolinemia, and hydroxyproline was detected only in some of the specimens. However, there was no problem in detecting the excessive urinary hydroxyproline. Unlike hyperprolinemia, there is no other amino acid abnormality in the urine.

The Presence of β-amino Acids

The β-amino acids stain poorly with ninhydrin reagent; the color development is enhanced by heating the chromatogram at 80 to 100°C for several minutes.

β-alanine is not found in normal urine. Its presence is easily detected by its location and by the blue color produced with ninhydrin stain. When it is present together with β-aminoisobutyric acid (BAIB), taurine and γ-aminobutyric acid (GABA), the most probable diagnosis is hyper-β-alaninemia. Isolated β-alaninuria has been reported in patients with tuberculosis[132] and as an abnormality related to transplantation rejection crisis in patients with renal allotransplants.[134] This amino acid was, however, not found in any of the renal transplanted patients studied by the author.

The Presence of Phosphoethanolamine

Phosphoethanolamine moves to a position near taurine; it runs slightly faster than taurine in HVE and slightly slower in BuAc (Figure 2-55). It may be mistaken for taurine if the latter is not present in the same specimen. It may be identified by the molybdate stain for phosphate. Phosphoethanolamine appears as a blue spot. This reagent cannot

*A 1 min volume of fresh urine is diluted to 2 ml and treated with 2 ml fresh alcoholic solution of 0.5% o-aminobenzaldehyde and 5% trichloroacetic acid. Wait for 30 min and centrifuge if any precipitate forms. The optical density is read at 443 nm using a millimolar extinction coefficient of 2.71 and corrected for urine blank. PC is unstable during refrigeration or freezing.

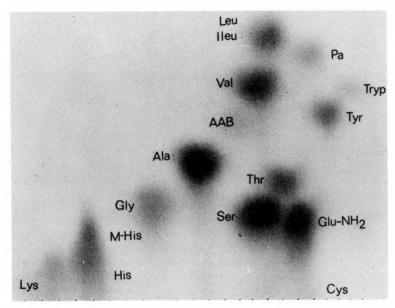

FIGURE 2-56. Two-dimensional paper chromatogram (HVE-BuAc) showing the characteristic urinary amino acid pattern in Hartnup disease, with the notable absence of proline and the relatively normal size of glycine spot.

be applied over ninhydrin and a separate two-dimensional chromatogram should be prepared. When this compound is found in the urine, the diagnosis of hypophosphatasia and pseudohypophosphatasia (the alkaline phosphatase activity is normal) should be considered.

Abnormalities in Alanine Concentration

Alanine is normally present in blood and urine. A spot of unusual intensity beneath proline on a one-dimensional chromatogram indicates an abnormal amount of alanine. Increased alanine concentration in the blood and urine has not been associated with a primary metabolic disorder. This phenomenon is secondary to altered pyruvate and lactate metabolism. When increased alanine is found in blood and urine, blood pyruvate and lactate should be measured. These abnormalities may be obvious only during episodes of clinical symptoms. Therefore, the diagnosis of pyruvate metabolic defects should not be dismissed simply because a normal value is obtained between attacks.

Low blood alanine concentrations have recently been found in ketotic hypoglycemia.[135] The difference is not evident with paper chromatography, but alanine can be conveniently measured quantitatively by the short column method.

Elevations of a Group of Amino Acids

The urinary amino acid pattern of Hartnup disease on a chromatogram is characteristic although it may at first sight resemble a generalized aminoaciduria. The neutral amino acids, serine, alanine, leucine, isoleucine, valine, phenylalanine, tyrosine, histidine, glutamine, threonine, and tryptophan, are greatly increased (Figure 2-56). Occasionally there may be a variation in the amount of these amino acids; some are not as prominent as others. Proline and hydroxyproline are notably absent and this is a *sine qua non* for the diagnosis of Hartnup disease. The glycine spot is usually of normal size and intensity or slightly increased. The basic amino acids and cystine are also present in normal amounts, but a mild increase in lysine is sometimes seen in the severely affected.

In Hartnup disease the transport of neutral amino acids is defective not only in the kidney but also in the intestine. Increased urinary excretion of the intestinal bacterial breakdown product of tryptophan (indican, indoleacetic acid, indolelactic acid, and indoleacetyl-glutamine) has been found in a number of patients. However, the absence of increased indole excretion does not negate the diagnosis. After oral tryptophan loading, urinary excretion of the indolic acids is increased and prolonged.[136,137] In contrast to normal individ-

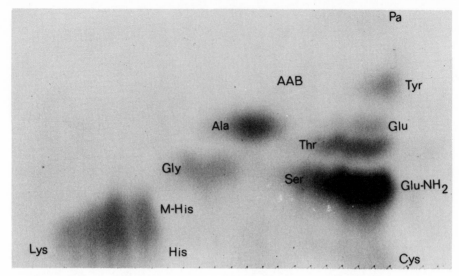

FIGURE 2-57. Two-dimensional paper chromatogram (HVE-BuAc) showing the urinary amino acid pattern in Reye's syndrome. Note the generalized hyperaminoaciduria with a particularly prominent glutamine spot.

uals who excrete only small amounts of indolic acids during the first 4 hr after loading, patients with Hartnup disease excrete much more and this excretion persists for 24 hr after loading. Kynurenine, being an endogenous metabolite of tryptophan, is excreted in greater quantity by normal individuals than by Hartnup patients. In some patients, stool amino acids* may be increased even without loading.[138]

Cystinuria, hyperdibasic aminoaciduria, and iminoglycinuria have been discussed in previous sections.

A generalized hyperaminoaciduria is sometimes present in liver disease and is usually found in Reye's syndrome. In the latter, glutamine, alanine, and lysine spots are quite prominent in the blood as well as in urine (Figure 2-57).

Blood and urine amino acid aberrations in various conditions other than inborn errors of amino acid metabolism have been extensively reviewed by Feigin.[145]

Generalized hyperaminoaciduria may be associated with various syndromes, metabolic errors of compounds other than amino acids, or exogenous causes. Examples of these are Lowe's syndrome, Fanconi syndrome, galactosemia, cystinosis, Wilson's disease, rickets, lead poisoning, ingestion of out-dated tetracycline, etc. (Figure 2-58). In some patients with rickets, the hyperaminoaciduria is limited to the quantitative increase of the usual amino acids seen in normal urine, namely alanine, glycine, histidine, threonine, serine, glutamine, and tyrosine. We have often referred to this pattern as "central core increase."

Cycloleucine, a chemical which has some antitumor effect, is known to cause a specific pattern of renal aminoaciduria resembling that of homozygote for cystinuria.[141]

Amino Acid Abnormalities in Miscellaneous Conditions

β-aminoisobutyric acid (BAIB) is sometimes mistaken for valine or methionine on onedimensional chromatograms. It has a slightly lower R_f than these two compounds. Transient excretion of BAIB may be found in conditions of increased tissue destruction such as irradiation, surgery, leukemia, and bladder tumor.

δ-aminolevulinic acid is near BAIB on twodimensional electrochromatogram. It reacts to form a yellow color in ninhydrin stain and is easily distinguished from other compounds. Its presence in the urine is often associated with acute lead poisoning.

*Stool is prepared for amino acid chromatography by homogenizing the specimen in 9 volumes of cold 10% isopropanol and diluting 5 ml of the homogenate to 20 ml with ethanol. After centrifugation 1/2 ml of the supernatant solution is spotted for two-dimensional electrochromatography.

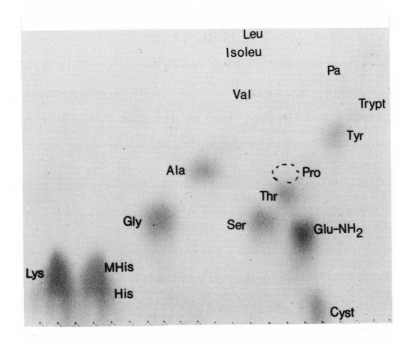

FIGURE 2-58. Two-dimensional paper chromatogram (HVE-BuAc) showing a generalized hyperaminoaciduria pattern in rickets.

In renal disease various abnormal amino acid patterns are seen. With renal tubular damage, a generalized aminoaciduria is common. Certain types of nephritis appear to be associated with hyperprolinemia, while with others mild lysine-cystinuria has been found.

Lysine-cystinuria has been reported in hereditary pancreatitis by one group of investigators,[142] but has not been confirmed by others. Since cystinuria is a common genetic trait,[131] it would not be surprising for heterozygosity for cystinuria and hereditary pancreatitis to coexist incidentally in the same individual.

Patients with severe burns excrete increased amounts of peptides. Two of those commonly seen on a two-dimensional chromatogram are shown in Figure 2-59. The spot, unknown 2, has also been found in patients with a variety of other diseases, such as metastatic neuroblastoma and pancreatic carcinoma.

Patients with active bone disease may excrete increased amounts of glycylproline and/or prolylhydroxyproline.[143,144] Bone collagen appears to be the most likely source of these dipeptides. Glycylproline is partially separated from alanine on the electrochromatogram (Figure 2-28), and with ninhydrin reagent it stains yellow initially and darkens to form a purple color within a few hours. Prolylhydroxyproline is near proline and

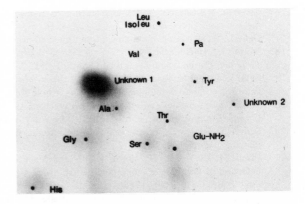

FIGURE 2-59. Two-dimensional paper chromatogram (HVE-BuAc) of urine from a patient with extensive burns, showing the presence of two unknown peptides.

stains yellow with ninhydrin. Both compounds are destroyed by hydrolysis.

Aspartylglycosamine appears as several yellow-brown spots on a two-dimensional chromatogram (HVE-BuAc) stained with ninhydrin reagent. The major spot is slow-moving at pH 2.0 (0.37 x the mobility of glycine) and has a low R_f of 0.09 in BuAc.[139] The urinary excretion of this compound has recently been found in two siblings with a deficiency of the enzyme responsible for its hydrolysis.[140]

REFERENCES

1. **Perry, T. L., Hansen, S., and MacDougall, L.,** Urinary screening tests in the prevention of mental deficiency, *Can. Med. Assoc. J.,* 95, 89, 1966.
2. **Medes, G.,** A new error of tyrosine metabolism: Tyrosinosis, the intermediary metabolism of tyrosine and phenylalanine, *Biochem. J.,* 26, 917, 1932.
3. **Spaeth, G. L. and Barber, G. W.,** Prevalence of homocystinuria among the mentally retarded: Evaluation of a specific screening test, *Pediatrics,* 40, 586, 1967.
4. **Sabater, J. and Maya, A.,** Specific spot test for homocystinuria in filter paper saturated with urine, *Clin. Chim. Acta,* 39, 261, 1972.
5. **Fiegl, F.,** *Spot Tests in Inorganic Analysis,* 5th ed., Elsevier, New York, 1958, 318.
6. **Conway, E. J.,** Apparatus for the microdetermination of certain volatile substances. IV. The blood ammonia, with observations on normal human blood, *Biochem. J.,* 29, 2755, 1935.
7. **Seligson, D. and Seligson, H.,** A microdiffusion method for the determination of nitrogen liberated as ammonia, *J. Lab. Clin. Med.,* 38, 324, 1951.
8. **Leffler, H. H.,** Measurement of ammonia in plasma, *Am. J. Clin. Pathol.,* 48, 233, 1967.
9. **Fenton, J. C. B. and Williams, A. H.,** Improved method for the estimation of plasma ammonia by ion exchange, *J. Clin. Pathol.,* 21, 14, 1968.
10. **Mondzac, A., Ehrlich, G. E., and Seegmiller, J. E.,** An enzymatic determination of ammonia in biological fluids, *J. Lab. Clin. Med.,* 66, 526, 1965.
11. **Rubin, M. and Knott, L.,** An enzymatic fluorometric method for ammonia determination, *Clin. Chim. Acta,* 18, 409, 1967.
12. **Levinson, S. A. and Macfate, R. P.,** Obermayer test, in *Clinical Laboratory Diagnosis,* Lea & Febiger, 1960, 559.
13. **Spencer, E. W., Ingram, V. M., and Levinthal, C.,** Electrophoresis: An accident and some precautions, *Science,* 152, 1722, 1966.
14. **Samuels, S.,** High-resolution screening of aminoacidurias, *Arch. Neurol.,* 10, 322, 1964.
15. **Mabry, C. C. and Todd, W. R.,** Quantitative measurement of individual and total free amino acids in urine, *J. Lab. Clin. Med.,* 61, 146, 1963.
16. **Consden, R., Gordon, A. H., and Martin, A. J. P.,** Qualitative analysis of proteins: A partition chromatographic method using paper, *Biochem. J.,* 38, 222, 1944.
17. **Dent, C. E.,** Detection of amino-acids in urine and other fluids, *Lancet,* 2, 637, 1946.
18. **Smith, I.,** Amino acids, amines, and related compounds, in *Chromatographic and Electrophoretic Techniques, Volume I, Chromatography,* 3rd ed., Smith, I., Ed., John Wiley & Sons, New York, 1969, 104.
19. **Shih, V. E. and Madigan, P. M.,** Improved paper-chromatographic method for imino acids. Incorporation of isatin, a color reagent, into developing solvents, *Clin. Chim. Acta,* 24, 481, 1969.
20. **Jepson, J. B.,** Indoles and related Ehrlich reactors, in *Chromatographic and Electrophoretic Techniques, Volume I, Chromatography,* 3rd ed., Smith, I., Ed., John Wiley & Sons, New York, 1969, 243.
21. **Scriver, C. R., Davies, E., and Cullen, A. M.,** Application of a simple method to the screening of plasma for a variety of aminoacidopathies, *Lancet,* 2, 230, 1964.
22. **Efron, M. L., Young, D., Moser, H. W., and MacCready, R. A.,** A simple chromatographic screening test for the detection of disorders of amino acid metabolism: A technic using whole blood or urine collected on filter paper, *N. Engl. J. Med.,* 270, 1378, 1964.
23. **Adriaenssens, K., Vanheule, R., and VanBelle, M.,** A new simple screening method for detecting pathological amino-acidemias with collection of blood on paper, *Clin. Chim. Acta,* 15, 362, 1967.
24. **Block, R. J., Dunum, E. L., and Zweig, G.,** Amino acids, amines and proteins, in *A Manual of Paper Chromatography and Paper Electrophoresis,* Academic Press, New York, 1955, 75.
25. **Berry, H. K., Leonard, C., Peters, H., Granger, M., and Chunekamrai, N.,** Detection of metabolic disorders. Chromatographic procedures and interpretation of results, *Clin. Chem.,* 14, 1033, 1968.
26. **Efron, M. L.,** High voltage paper electrophoresis, in *Chromatographic and Electrophoretic Techniques, Volume II, Zone Electrophoresis,* 3rd ed., Smith, I., Ed., John Wiley & Sons, New York, 1968, 166.
27. **Smith, I., Rider, L. J., and Lerner, R. P.,** A rapid two-way screening procedure for whole blood and urine, *Biochem. Med.,* 1, 9, 1967.
28. **Smith, I.,** Paper chromatographic apparatus and techniques, in *Chromatographic and Electrophoretic Techniques, Volume I, Chromatography,* 3rd ed., Smith, I., Ed., John Wiley & Sons, New York, 1969, 7.
29. **Smith, I., Rider, J., and Lerner, R. P.,** Chromatography of amino acids, indoles and imidazoles on thin layers of Avicel and cellulose and on paper, *J. Chromatogr.,* 26, 449, 1967.
30. **Stahl, E.,** Instruments used in thin-layer chromatography and their operation, in *Thin-Layer Chromatography: A Laboratory Handbook,* Stahl, E., Ed., Academic Press, New York, 1965, 5.
31. **Kraffczyk, F., Helger, R., and Lang, H.,** Two-dimensional thin-layer chromatography on two-layer plates of amino acids, *Clin. Chem.,* 16, 662, 1970.
32. **Ersser, R. S. and Seakins, J. W. T.,** Screening for aminoacidopathies with prepared cellulose layers on aluminum foil, *Nature,* 223, 1388, 1969.

33. **Bremer, H. J., Nützenadel, W., and Bickel, H.,** Dünnschichtchromatographische Methoden zur Erkennung von Amino-acidurien, Aminoacidaemien Sowie der Galactosaemie, *Mschr. f. Kinderheilkunde,* 117, 32, 1969.

34. **Plöchl, E.,** Dünnschichtchromatographische auftrennung von Blutaminosauren aus mit Blut beschickten Filter-papier-Plättchen, *Clin. Chim. Acta,* 21, 271, 1968.

35. **Culley, W. J.,** A rapid and simple thin-layer chromatographic method for amino acids in blood, *Clin. Chem.,* 15, 902, 1969.

36. **Wadman, S. K., Fabery de Jonge, H., and deBree, P. K.,** Rapid, high-resolution, two-dimensional amino acid chromatography on micro scale chromatograms, *Clin. Chim. Acta,* 25, 87, 1969.

37. **White, H. H.,** Separation of amino acids in physiological fluids by two-dimensional thin-layer chromatography, *Clin. Chim. Acta,* 21, 297, 1968.

38. **Kraffczyk, F., Helger, R., Lang, H., and Bremer, H. J.,** Thin layer chromatographic screening test for amino acid anomalies in urine without desalting using internal standards, *Clin. Chim. Acta,* 35, 345, 1971.

39. **Peters, J. H., Lin, S. C., Berridge, B. J., Cummings, J. G., and Chao, W. R.,** Amino acids, including asparagine and glutamine, in plasma and urine of normal human subjects, *Proc. Soc. Exp. Biol. Med.,* 131, 281, 1969.

40. **Scriver, C. R., Clow, C. L., and Lamm, P.,** Plasma amino acids: screening, quantitation and interpretation, *Am. J. Clin. Nutr.,* 24, 876, 1971.

41. **Perry, T. L. and Hansen, S.,** Technical pitfalls leading to errors in the quantitation of plasma amino acids, *Clin. Chim. Acta,* 25, 53, 1969.

42. **Stein, W. H. and Moore, S.,** The free amino acids of human blood plasma, *J. Biol. Chem.,* 211, 915, 1954.

43. **Gerritsen, T., Rehberg, M. L., and Waisman, H. A.,** On the determination of free amino acids in serum, *Anal. Biochem.,* 11, 460, 1965.

44. **Dickinson, J. C., Rosenblum, H., and Hamilton, P. B.,** Ion exchange chromatography of the free amino acids in the plasma of the newborn infant, *Pediatrics,* 36, 2, 1965.

45. **DeWolfe, M. S., Baskurt, S., and Cochrane, W. A.,** Automatic amino acid analysis of blood serum and plasma, *Clin. Biochem.,* 1, 75, 1967.

46. **Hamilton, P. B.,** Ion exchange chromatography of amino acids — microdetermination of free amino acids in serum, *Ann. N.Y. Acad. Sci.,* 102, 55, 1962.

47. **Levy, H. L. and Barkin, E.,** Comparison of amino acid concentrations between plasma and erythrocytes. Studies in normal human subjects and those with metabolic disorders, *J. Lab. Clin. Med.,* 78, 517, 1971.

48. **Armstrong, M. D. and Stemmermann, M. G.,** An occurrence of argininosuccinic aciduria, *Pediatrics,* 33, 280, 1964.

49. **Perry, T. L., Stedman, D., and Hansen, S.,** A versatile lithium buffer elution system for single column automatic amino acid chromatography, *J. Chromatogr.,* 38, 460, 1968.

50. **Benson, J. V., Jr., Gordon, M. J., and Patterson, J. A.,** Accelerated chromatographic analysis of amino acids in physiological fluids containing glutamine and asparagine, *Anal. Biochem.,* 18, 228, 1967.

51. **Peters, J. H., Berridge, B. J., Jr., Cummings, J. G., and Lin, S. C.,** Column chromatographic analysis of neutral and acidic amino acids using lithium buffers, *Anal. Biochem.,* 23, 459, 1968.

52. **Kedenburg, C. P.,** A lithium buffer system for accelerated single-column amino acid analysis in physiological fluids, *Anal. Biochem.,* 40, 35, 1971.

53. **Moore, S. and Stein, W. H.,** Chromatography of amino acids on sulfonated polystyrene resins, *J. Biol. Chem.,* 192, 663, 1951.

54. **Piez, K. A. and Morris, L.,** A modified procedure for the automatic analysis of amino acids, *Anal. Biochem.,* 1, 187, 1960.

55. **Efron, M. L.,** Quantitative estimation of amino acids in physiological fluids using a Technicon amino acid analyzer: A modified technique with improved separation of amino acids and a simplified method for preparation of blood samples, in *Automation in Analytical Chemistry (Technicon Symposia, 1965),* Skeggs, L. T., Jr., Ed., Mediad, Inc., New York, 1966, 637.

56. **Moore, S. and Stein, W. H.,** Procedures for the chromatographic determination of amino acids on four per cent cross-linked sulfonated polystyrene resins, *J. Biol. Chem.,* 211, 893, 1954.

57. **Hamilton, P. B.,** Ion exchange chromatography of amino acids, *Anal. Chem.,* 35, 2055, 1963.

58. **Zacharius, R. M. and Talley, E. A.,** Elution behavior of naturally occurring ninhydrin-positive compounds during ion-exchange chromatography, *Anal. Chem.,* 34, 1551, 1962.

59. **Scriver, C. R., Davies, E., and Lamm, P.,** Accelerated selective short column chromatography of neutral and acidic amino acids on a Beckman-Spinco Analyzer, modified for simultaneous analysis of two samples, *Clin. Biochem.,* 1, 179, 1968.

60. **Shih, V. E., Efron, M. L., and Mechanic, G. L.,** Rapid short-column chromatography of amino acids: A method for blood and urine specimens in the diagnosis and treatment of metabolic disease, *Anal. Biochem.,* 20, 299, 1967.

61. **Benson, J. V., Jr., Cormick, J., and Patterson, J. A.,** Accelerated chromatography of amino acids associated with phenylketonuria, leucinosis (maple syrup urine disease), and other inborn errors of metabolism, *Anal. Biol.,* 18, 481, 1967.

62. **Mechanic, G., Efron, M. L., and Shih, V. E.,** A rapid quantitative estimation of tyrosine and phenylalanine by ion-exchange chromatography, *Anal. Biochem.,* 16, 420, 1966.

63. **DeMarco, C., Mosti, R., and Cavallini, D.,** Column chromatography of some sulfur-containing amino acids, *J. Chromatogr.,* 18, 492, 1965.
64. **DeMarco, C., Coletta, M., and Cavallini, D.,** Column chromatography of phosphoserine, phosphoethanolamine and S-sulfoglutathione and their identification in the presence of other amino acids, *J. Chromatogr.,* 20, 500, 1965.
65. **Ratner, S. and Kunkemueller, M.,** Separation and properties of argininosuccinate and its two anhydrides and their detection in biological materials, *Biochemistry,* 5, 1821, 1966.
66. **Barber, G. W.,** Multiple autoanalyzer manifolds for sulfur amino acid chromatography, in *Automation in Analytical Chemistry (Technicon Symposia, 1966),* Vol. 1. Mediad, Inc., New York, 1967, 401.
67. **Shih, V. E., Jones, T. C., Levy, H. L., and Madigan, P. M.,** Arginase deficiency in Masaca fascicularis. I. Arginase activity and arginine concentration in erythrocytes and in liver, *Pediatr. Res.,* 6, 548, 1972.
68. **Peters, J. H. and Berridge, B. J., Jr.,** The determination of amino acids in plasma and urine by ion-exchange chromatography, *Chromatogr. Rev.,* 12, 157, 1970.
69. **Dickinson, J. C., Rosenblum, H., and Hamilton, P. B.,** Ion exchange chromatography of the free amino acids in the plasma of infants under 2,500 gm at birth, *Pediatrics,* 45, 606, 1970.
70. **Snyderman, S. E., Holt, L. E., Jr., Norton, P. M., Roitman, E., and Phansalkar, S. V.,** The plasma aminogram. I. Influence of the level of protein intake and a comparison of whole protein and amino acid diets, *Pediatr. Res.,* 2, 131, 1968.
71. **Soupart, P.,** Free amino acids of blood and urine in the human in *Amino Acid Pools. Distribution, Formation and Function of Free Amino Acids,* Holden, J. T., Ed., Elsevier, New York, 1962, 220.
72. **Scriver, C. R. and Davies, E.,** Endogenous renal clearance rates of free amino acids in pre-pubertal children, *Pediatrics,* 32, 395, 1968.
73. **Nyhan, W. L., Yujnovsky, A. O., and Wehr, R. R.,** Amino acids and cell growth, in *Human Growth. Body Composition, Cell Growth, Energy, and Intelligence,* Cheek, D. B., Ed., Lea & Febiger, Philadelphia, 1968, 396.
74. **Scriver, C. R., Clow, C. L., and Lamm, P.,** Plasma amino acids: screening, quantitation, and interpretation, *Am. J. Clin. Nutr.,* 24, 876, 1971.
75. **Ackermann, P. G. and Kheim, T.,** Plasma amino acids in young and older adult human subjects, *Clin. Chem.,* 10, 32, 1964.
76. **Levy, H. L., Shih, V. E., Madigan, P. M., Karolkewicz, V., Carr, J. R., Lum, A., Richards, A. A., Crawford, J. D., and MacCready, R. A.,** Hypermethioninemia with other hyperaminoacidemias, *Am. J. Dis. Child.,* 117, 96, 1969.
77. **Hamilton, P. B.,** High resolution ion exchange chromatography of amino acids: The analysis of urine and other biological fluids, in *Automation in Analytical Chemistry (Technicon Symposia, 1967),* Vol. 1, Mediad, Inc., New York, 1968, 317.
78. **Armstrong, M. D., Yates, K. N., and Connelly, J. P.,** Amino acid excretion of newborn infants during the first twenty-four hours of life, *Pediatrics,* 33, 975, 1964.
79. **Levy, H. L., Mudd, S. H., Schulman, J. D., Dreyfus, P. M., and Abeles, R. H.,** A derangement of B_{12} metabolism associated with homocystinemia, cystathioninemia, hypomethioninemia and methylmalonic aciduria, *Am. J. Med.,* 48, 390, 1970.
80. **Carver, M. J. and Paska, R.,** Ion-exchange chromatography of urinary amino acids, *Clin. Chim. Acta,* 6, 721, 1961.
81. **Scott-Emuakpor, A., Higgins, J. V., and Kohrman, A. F.,** Citrullinemia: A new case, with implications concerning adaptation to defective urea synthesis, *Pediatr. Res.,* 6, 626, 1972.
82. **Stein, W. H.,** A chromatographic investigation of the amino acid constituents of normal urine, *J. Biol. Chem.,* 201, 45, 1953.
83. **Logan, A., Schlicke, C. P., and Manning, G. B.,** Familial pancreatitis, *Am. J. Surg.,* 115, 112, 1968.
84. **Berridge, B. J., Jr., Chao, W. R., and Peters, J. H.,** Analysis of plasma and urinary amino acids by ion-exchange column chromatography, *Am. J. Clin. Nutr.,* 24, 934, 1971.
85. **Perry, T. L., Diamond, S., Hansen, S., and Stedman, D.,** Plasma-aminoacid levels in Huntington's Chorea, *Lancet,* 1, 806, 1969.
86. **Dickinson, J. C. and Hamilton, P. B.,** The free amino acids of human spinal fluid determined by ion exchange chromatography, *J. Neurochem.,* 13, 1179, 1966.
87. **Van Sande, M., Mardens, Y., Adriaenssens, K., and Lowenthal, A.,** The free amino acids in human cerebrospinal fluid, *J. Neurochem.,* 17, 125, 1970.
88. **Guthrie, R. and Susi, A.,** A simple phenylalanine method for detecting phenylketonuria in large populations of newborn infants, *Pediatrics,* 32, 338, 1963.
89. **Guthrie, R.,** Screening for "inborn errors of metabolism" in the newborn infant – a multiple test program, *Birth Defects Orig. Art. Ser.,* 4, 92, 1968.
90. **Bolinder, A. E.,** Microbiological plate assay methods for amino acids, *Nutr. Rep. Int.,* 6, 67, 1972.
91. **Murphey, W. H., Patchen, L., and Guthrie, R.,** Screening tests for argininosuccinic aciduria, orotic aciduria, and other inherited enzyme deficiencies using dried blood specimens, *Biochem. Genet.,* 6, 51, 1972.
92. **Levy, H. L., Baullinger, P. C., and Madigan, P. M.,** A rapid procedure for the determination of phenylalanine and tyrosine from blood filter paper specimens, *Clin. Chim. Acta,* 31, 447, 1971.
93. **McCaman, M. and Robins, E.,** Fluorometric method for the determination of phenylalanine in serum, *J. Lab. Clin. Med.,* 59, 885, 1962.

94. **Hill, J. B., Summer, G. K., Pender, M. W., and Roszel, N. O.,** An automated procedure for blood phenylalanine, *Clin. Chem.,* 11, 451, 1965.

95. **Ambrose, J. A.,** A shortened method for the fluorometric determination of phenylalanine, *Clin. Chem.,* 15, 15, 1969.

96. **Waalkes, T. P. and Udenfriend, S.,** A fluorometric method for the estimation of tyrosine in plasma and tissues, *J. Lab. Clin. Med.,* 50, 733, 1957.

97. **Ambrose, J. A., Sullivan, P., Ingerson, A., and Brown, R. L.,** Fluorometric determination of tyrosine, *Clin. Chem.,* 15, 611, 1969.

98. **Ambrose, J. A., Crimm, A., Burton, J., Paullin, K., and Ross, C.,** Fluorometric determination of histidine, *Clin. Chem.,* 15, 361, 1969.

99. **Fisch, R. O., Anthony, B. F., Bauer, H., and Bruhl, H. H.,** The effect of antibiotics on the results of the Guthrie test given to phenylketonuric patients, *J. Pediatr.,* 73, 685, 1968.

100. **Koch, R., Williamson, M. L., Donnell, G. N., Guthrie, R., Straus, R., Coffelt, R. W., and Fish, C. H.,** A cooperative study of two methods for phenylalanine determination: McCaman-Robins fluorimetric and microbiologic inhibition methods, *J. Pediatr.,* 68, 905, 1966.

101. **Bixby, E. M., Pallatao, L. G., and Pryles, C. V.,** Evaluation of the Bacillus subtilis inhibitional assay technic as a screening procedure for the detection of phenylketonuria, *N. Engl. J. Med.,* 268, 648, 1963.

102. **Ambrose, J. A.,** Evaluation studies resulting in the standardization of the bacterial inhibition assay for blood phenylalanine, *Health Lab. Sci.,* 6, 199, 1969.

103. **Efron, M. L., McPherson, T. C., Shih, V. E., Welsh, C. F., and MacCready, R. A.,** D-Methioninuria due to DL-methionine ingestion, *Am. J. Dis. Child.,* 117, 104, 1969.

104. **Steqink, L. D., Schmitt, J. L., Meyer, P. D., and Kain, P. H.,** Effect of diets fortified with DL-methionine on urinary and plasma methionine levels in young infants, *J. Pediatr.,* 79, 648, 1971.

105. **Perry, T. L., Hansen, S., Hestrin, M., and MacIntyre, L.,** Exogenous urinary amines of plant origin, *Clin. Chim. Acta,* 11, 24, 1965.

106. **Meilman, E., Uriretzky, M. M., and Rapoport, C. M.,** Urinary hydroxyproline peptides, *J. Clin. Invest.,* 42, 40, 1963.

107. **Perry, T. L., Dixon, G. H., and Hansen, S.,** Iatrogenic urinary amino-acid derived from penicillin, *Nature,* 206, 895, 1965.

108. **Scriver, C. R. and Perry, T. L.,** Disorders of β-alanine and carnosine metabolism, in *The Metabolic Basis of Inherited Disease,* 3rd ed., Stanbury, J. B., Wyngaarden, J. B., and Fredrickson, D. S., Eds., McGraw-Hill, New York, 1972, 476.

109. **Hyanek, J., Bremer, H. J., and Slavik, M.,** "Homocystinuria" and urinary excretion of β-amino acids in patients treated with 6-azauridine, *Clin. Chim. Acta,* 25, 288, 1969.

110. **Luhby, A. L. and Cooperman, J. M.,** Aminoimidazolecarboxamide excretion in vitamin B_{12} and folic-acid deficiencies, *Am. J. Clin. Nutr.,* 7, 397, 1959.

111. **Shih, V. E. and Schulman, J. D.,** N-Acetylcysteine-cysteine disulfide excretion in the urine following N-acetylcysteine administration, *J. Pediatr.,* 74, 129, 1969.

112. **Levy, H. L., Madigan, P. M., and Lum, A.,** Fecal contamination in urine amino acid screening: Artifactual cause of hyperaminoaciduria, *Am. J. Clin. Pathol.,* 51, 765, 1969.

113. **Gjessing, L. R. and Lindeman, R.,** p-Hydroxyphenylacetic acid in cystic fibrosis, *Lancet,* 2, 47, 1967.

114. **Kang, E. S. and Gerald, P. S.,** Alcaptonuria: Rapid semiquantitative determination of homogentisic acid in urine, *J. Pediatr.,* 76, 393, 1970.

115. **Libit, S. A., Ulstrom, R. A., and Doeden, D.,** Fecal pseudomonas aeruginosa as a cause of the blue diaper syndrome, *J. Pediatr.,* 81, 546, 1972.

116. **Buist, N. R. M., Ramberg, D. A., Strandholm, J. J., and Ferry, P. C.,** Copper poppers: a benign cause of blue diapers, *Arch. Dis. Child.,* 46, 873, 1971.

117. **Armstrong, M. D., Shaw, K. N. F., Gortatowski, M. J., and Singer, H.,** The indole acids of human urine. Paper chromatography of indole acids, *J. Biol. Chem.,* 232, 17, 1958.

118. **Thalhammer, O., Scheibenreiter, S., and Pantiltschko, M.,** Histidinemia: detection by routine newborn screening and biochemical observations on three unrelated cases, *Z. Kinderheilkd.,* 109, 279, 1971.

119. **Levy, H. L., Baden, H. P., and Shih, V. E.,** A simple indirect method of detecting the enzyme defect in histidinemia, *J. Pediatr.,* 75, 1056, 1969.

120. **Perry, T. L., Hansen, S., and Love, D. L.,** Serum carnosinase deficiency in carnosinaemia, *Lancet,* 1, 1229, 1968.

121. **Van Munster, P. J. J., Trijbels, J. M. F., Van Heeswijk, P. J., Schut-Jansen, B., and Moerkerk, C.,** A new sensitive method for the determination of serum carnosinase activity using L-carnosine-(I-^{14}C) β-alanyl as substrate, *Clin. Chim. Acta,* 29, 243, 1970.

122. **Gjessing, L. R.,** Studies of functional neural tumors. II. Cystathioninuria, *Scand. J. Clin. Lab. Invest.,* 15, 474, 1963.

123. **Geiser, C. F. and Efron, M. L.,** Cystathioninuria in patients with neuroblastoma or ganglioneuroblastoma: Its correlation to vanilmandelic acid excretion and its value in diagnosis and therapy, *Cancer,* 22, 856, 1968.

124. **Geiser, C. F., Baez, A., Schindler, A. M., and Shih, V. E.,** Epithelial hepatoblastoma associated with congenital hemihypertrophy and cystathioninuria: Presentation of a case, *Pediatrics,* 46, 66, 1970.

125. Shaw, K. N. F., Lieberman, E., Koch, R., and Donnell, G. N., Cystathioninuria, *Am. J. Dis. Child.,* 113, 119, 1967.
126. Felig, P., Owen, O. E., Wahren, J., and Cahill, G. F., Jr., Amino acid metabolism during prolonged starvation, *J. Clin. Invest.,* 48, 584, 1969.
127. Tomlinson, S. and Westall, R. G., Argininosuccinic aciduria: Argininosuccinase and arginase in human blood cells, *Clin. Sci.,* 26, 261, 1964.
128. Shih, V. E., Littlefield, J. W., and Moser, H. W., Argininosuccinase deficiency in fibroblasts cultured from patients with argininosuccinic aciduria, *Biochem. Genet.,* 3, 81, 1969.
129. Gatfield, P. D., Tallen, E., Hinton, G. G., Wallace, A. C., Abdelnour, G. M., and Haust, M. D., Hyperpipecolatemia: A new metabolic disorder associated with neuropathy and hepatomegaly, *Can. Med. Assoc. J.,* 99, 1215, 1968.
130. Pasieka, A. E. and Morgan, J. F., Specific determination of proline in biologic materials, *Proc. Soc. Exp. Biol. Med.,* 93, 54, 1956.
131. Levy, H. L., Madigan, P. M., and Shih, V. E., Massachusetts metabolic disorders screening program. I. Technics and results of urine screening, *Pediatrics,* 49, 825, 1972.
132. Efron, M. L., Familial hyperprolinemia: Report of a second case associated with congenital renal malformations, hereditary hematuria, and mild mental retardation, with demonstration of an enzyme defect, *N. Engl. J. Med.,* 272, 1243, 1965.
133. Takao, T., Yasumitsu, T., Uozumi, T., Kakimoto, Y., and Kanazawa, A., β-alaninuria in patients with tuberculosis, *Nature,* 217, 365, 1968.
134. Gras, G., Tuset, N., Caralps, A., Gil-Varnet, J. M., Margrina, N., Brulles, A., and Conde, M., β-Alaninuria following human renal allotransplantation, *Clin. Chim. Acta,* 20, 295, 1968.
135. Pagliara, A. S., Karl, I. E., De Vivo, D. C., Feigin, R. D., and Kipnis, D. M., Hypoalaninemia: a concomitant of ketotic hypoglycemia, *J. Clin. Invest.,* 51, 1440, 1972.
136. Milne, M. D., Crawford, M. A., Girao, C. B., and Loughridge, L., The metabolic disorders in Hartnup disease, *Quart. J. Med.,* 29, 407, 1960.
137. Shih, V. E., Bixby, E. M., Alpers, D. H., Bartsocas, C. S., and Thier, S. O., Studies of intestinal transport defect in Hartnup disease, *Gastroenterology,* 61, 445, 1971.
138. Scriver, C. R., Hartnup disease. A genetic modification of intestinal and renal transport of certain neutral alpha-amino acids, *N. Engl. J. Med.,* 273, 530, 1965.
139. Jenner, F. A. and Pollitt, R. J., Large quantities of 2-acetamido-1 (β-L-aspartamido)-1,2-dideoxyglucose in the urine of mentally retarded siblings, *Biochem. J.,* 103, 48, 1967.
140. Pollitt, R. J., Jenner, F. A., and Merskey, H., Aspartylglycosaminuria: An inborn error of metabolism associated with mental defect, *Lancet,* 2, 253, 1968.
141. Brown, R. R., Aminoaciduria resulting from cycloleucine administration in man, *Science,* 157, 432, 1967.
142. Gross, J. B., Ulrich, J. A., and Maher, F. T., Further observations on the hereditary form of pancreatitis, in *Ciba Foundation Symposium on the Exocrine Pancreas: Normal and Abnormal Functions,* deReuck, A. B. S. and Cameron, M. P., Eds., Little, Brown and Company, Boston, 1961, 278.
143. Seakins, J. W. T., Peptiduria in an unusual bone disorder: isolation of two peptides, *Arch. Dis. Child.,* 38, 215, 1963.
144. Scriver, C. R., Glycyl-proline in urine of humans with bone diseases, *Can. J. Physiol. Pharmacol.,* 42, 357, 1964.
145. Feigin, R. D., Blood and urine amino acid aberrations: Physiologic and pathological changes in patients without inborn errors of amino acid metabolism, *Am. J. Dis. Child.,* 117, 24, 1969.

Chapter 3

SUGARS

INTRODUCTION

Hexoses (six-carbon sugars) are important compounds in metabolism. Of these, glucose is the most important sugar; it is a source of energy for other metabolic processes. Several pentoses (five-carbon sugars) are present in humans; they are either intermediates in metabolism or constituents of nucleic acid. Dietary sugars consist of glucose, fructose, arabinose, xylose, sucrose (glucose-fructose disaccharide), and lactose (glucose-galactose disaccharide). The disaccharides are hydrolyzed to monosaccharides in the intestine and then absorbed. Fructose and galactose are converted to glucose in the liver, whereas arabinose and xylose are not metabolizable.

Table 3–1 shows the sugars that have been found in the urine and which are detectable by the following methods.

SCREENING TESTS FOR SUGARS

Chemical Tests for Reducing Substances

A simple modification of the Benedict test made available by a commercial source is now considered a standard laboratory procedure.

Reagent: A commercially available reagent tablet* containing copper sulfate, caustic soda, sodium carbonate, and citric acid.

Method: Place 5 drops of urine and 10 drops of water in a test tube and add to it the reagent tablet. Do not shake test tube until 15 sec after the bubbling reaction has stopped.

Results and interpretation: Green to orange colors (1+ to 4+) indicate the presence of various amounts of reducing substances. A color chart for comparison is included in the package insert.

Reducing sugars such as glucose, galactose, fructose, lactose, mannose, and xylulose, and such reducing substances as sialic acid and homogentisic acid give a positive reaction. Sucrose is not a reducing sugar and gives a negative reaction. Differentiation among these compounds may be achieved by chromatographic separation, reaction with specific oxidases, etc.

Urine from patients receiving cephalothin** and ampicillin gives a dark brown color which must not be mistaken as a positive reaction.

TABLE 3-1

Sugars Which Have Been Found in Urine

		Aldose	Ketose	Reducing property
Monosaccharides	Hexoses	Glucose		+
		Galactose		+
			Fructose	+
	Pentoses	Xylose		+
			Xylulose	+
		Ribose		+
			Ribulose	+
		Arabinose		+
Disaccharides	Lactose			+
	Sucrose			−

*Clinitest®, Ames Company
**Keflin®, Eli Lilly and Company

Anthrone Test for Carbohydrates[1]

Reagent: Anthrone 0.2% (w/v) in concentrated sulfuric acid.

Method: To 2 ml of the reagent add 0.2 ml of urine and mix. Berry et al.[2] have modified the method and used it as a spot test. To 3 drops of urine in a white porcelain hanging dish add 12 drops of the anthrone reagent and mix with a glass stirring rod.

Results and interpretation: This is a sensitive but nonspecific test. A green to blue-green color is a positive reaction, indicating the presence of any carbohydrates. Both reducing and nonreducing sugars such as sucrose will react. Other carbohydrates such as glycogen and cellulose (filter paper) also give positive results. Therefore, a positive result should be followed by more specific tests to determine the nature of the carbohydrate.

Filter paper specimens cannot be tested because filter paper itself reacts with the reagent. Specimens can be first eluted by soaking the filter paper in 3 drops of water in the hanging dish for 10 min. Remove the disc before adding the anthrone reagent.

Glucose Oxidase Test

Reagent: A commercially available test strip* is a colorimetric test based on the enzymatic action of glucose oxidase. It is a widely used simple test for semiquantitative estimation of urinary glucose.

Method and results: The test strip is dipped into the urine and the color change is compared using a color chart. The presence of ascorbic acid, bilirubin glucuronide, or homogentisic acid may inhibit this enzyme activity and cause a false-negative result. Glycosuria is present in diabetes mellitus, Fanconi syndrome, and renal glycosuria. Compounds toxic to renal tubules may cause glycosuria.

Since the glucose oxidase test strip is specific for glucose and will not detect other important reducing sugars, it is not a good screening test. The test for all reducing substances (Clinitest) rather than that for glucose only should be used as a routine screening procedure.

for testing glucose in filter paper urine specimens (similar to that described below for galactosemia).

Galactose Oxidase Test

A commercially available kit** can be adapted galactose oxidase is used for blood galactose determination. The activity of the galactose oxidase is coupled with that of a peroxidase, and a chromogen is oxidized giving a blue color which is measured by a spectrophotometer. This kit has been adapted by Woolf[3] and by Sabater**** to be used as a spot screening test for the presence of galactose in the urine.

Reagents: The kit consists of two vials, one containing the enzymes, galactose oxidase and peroxidase, and buffer, and one containing the chromogen. The contents are diluted as follows: To the enzyme and buffer vial, add 2.5 ml distilled water (Reagent A); to the chromogen vial add 1 ml ethanol and 1.5 ml distilled water (Reagent B). Both reagents are stable for one month at $4°C$.

Method and results: A 3/8 in. disc of filter paper urine specimen is placed in a white porcelain hanging drop dish. The disc is first wet with 1 drop of Reagent A, and 5 min later with 1 drop of Reagent B. The development of a blue color is positive. In addition to galactose, lactose reacts to give a positive reaction.

Microbiological Tests for Galactose

The microbiological assay for galactose is recommended only for routine mass screening, particularly newborn screening. In other age groups or for a small number of specimens, a test for the presence of reducing substances in urine is much more useful and practical.

The basic equipment needed for these tests is similar to that used in microbiological tests for amino acids.

Metabolite Inhibition Assay

This assay was developed by Dr. Guthrie***** using an *E. coli* mutant (W-3101) lacking Gal-1-P uridyltransferase which is sensitive to galactose, Gal-1-P, and valine; the presence of any of these compounds inhibits the growth of this mutant. Therefore, a zone of inhibition is a positive result.

---

*Clinistix® and Labstix®, Ames Co., and Tes-tape®, Eli Lilly & Co.
**Glucostat®, Worthington Biochemical Co.
***Galactostat®, Worthington Biochemical Co.
****Dr. Juan Sabater, Barcelona, Spain, personal communication, 1968.
*****Dr. Robert Guthrie, State University of New York at Buffalo, N. Y., personal communication.

Preparation of Culture Medium

The composition of the culture medium is as follows:

	Amount/l. assay medium
Glycerol	20.0 g
K_2HPO_4	6.0 g
KH_2PO_4	2.0 g
NH_4Cl	5.0 g
NH_4NO_2	1.0 g
Na_2SO_4	2.0 g
$MgSO_4 \cdot 7H_2O$	0.1 g
$CaCl_2$	20.0 mg
Trace element solution	1.0 ml

($(NH_4)_2MoO_4$ – 40.8 mg; H_3BO_3 – 57.0 mg; $FeCl_3 \cdot 6H_2O$ – 935.0 mg; $CuSO_4 \cdot 5H_2O$ – 393.0 mg; $MnCl_2 \cdot 4H_2O$ – 72.0 mg; $ZnCl_2$ – 4.16 mg in 1 l.)

Glycerol is prepared as a 40% solution and the phosphate buffer is prepared 20 times concentrated; each is sterilized separately and added to the remainder of the medium before use. The other components are dissolved in water, sterilized, and then mixed with glycerol and phosphate buffer to make the concentrations twice those indicated above.

Preparation of Inoculum

A cell suspension of *E. coli* mutant W-3101 is prepared to give an O.D.$_{550}$ of 1.0 to 1.1. It should be cautioned that this organism is a rather unstable mutant and can lose its galactose sensitivity during cultivation. Therefore, cells should not be subcultured indefinitely.

Preparation of Agar Test Plate

Three grams of agar are dissolved in 100 ml water by placing the container in a boiling water bath. The agar bottle is then cooled to 50 to 55°C. The culture medium prepared as described above is warmed to 50 to 55°C. Thoroughly mix equal volumes of the culture medium and the 3% agar, and 4 ml of the cell suspension per 200 ml of the mixture (or per tray 7 in. x 11 in. in size) by pouring back and forth. After the plates are hardened, they are ready for use.

Collection and Preparation of Blood Specimens and Standards

Blood specimens collected as for routine screening are impregnated on filter paper S.S. #903. Standards are prepared by adding known amounts of galactose to a blood specimen at 5, 10, 30, and 100 mg/100 ml, and spotted on filter paper.

Procedure

Discs of unknown blood specimens and graded standards 1/8 in. in diameter are placed in rows 1.5 in. apart on the agar plate and incubated at 37°C overnight.

Paigen Bacteriophage Test

This is an improved assay for galactose recently developed by Paigen.* This test utilizes a mutant strain of *E. coli* Q396 which is resistant to destruction by bacteriophage in the presence of galactose or lactose. Thus, there is a zone of growth around the discs of blood containing galactose. This phage test is better than the metabolite inhibition assay simply because the zone of growth is easier to read than a zone of inhibition, and mutant *E. coli* Q396 is more stable than mutant *E. coli* W-3101 used in the metabolite inhibition assay.

Preparation of Assay Medium

Base medium: This is prepared by mixing the following stock solutions in the proportions indicated:

Items	Stock solution (g/100 ml)	Proportion (ml)
$FeCl_3 \cdot 6H_2O$	0.005 (in 0.1 N HCl)	1.0
Gelatin	0.1	1.0
NH_4Cl	10.0	1.0
$MgSO_4 \cdot 7H_2O$	7.5	1.0
NaCl	5.0	1.0
L-glycine	10.0	1.0
Monosodium glutamate	0.2	1.0
L-methionine	0.2	1.0
L-threonine	0.2	1.0
Glycerol	20.0	5.0
Tris buffer, 1 M, pH 8.0		10.0

*Dr. Kenneth Paigen, Roswell Park Memorial Institute, Buffalo, N. Y., personal communication.

This base medium can be mixed, autoclaved, and stored indefinitely at 4°C.

Growth medium: The base medium is mixed with the following components in the proportions indicated:

	ml
Base medium	25.0
$CaCl_2 \cdot 2H_2O$, 0.3 g/100 ml	1.0
Thiamine, 0.05 mg/ml	1.0
KH_2PO_4, 1.25 g/100 ml	1.0
Water, to	100.0 ml

Preparation of Silicate

Sodium silicate, 7.5%: Dilute silicate of soda to 24 g of total solids/100 ml by mixing 40 ml of commercial silicate of soda grade 40 (total solids: 56.5 g/100 ml) with 60 ml water (specific gravity of 1.16 to 1.17). This solution is then purified by passing through a column of activated charcoal.* Dilute to 7.5% with water.

Preparation of the Inoculum and Phage

E. coli Q396 cells are supplied frozen in ampules containing 1.0 ml cells each.** Add one vial to 0.1 to 3.0 l. of fresh growth medium and aerate vigorously at 37°C until the O.D.$_{550}$ rises to about 2.0. These cells can be stored at room temperature for a maximum of four or five weeks.

Phage C21 is supplied in suspension medium at a titer of 1.25×10^8 phages/ml.**

Collection of Blood Specimens and Preparation of Standards

Blood collected on filter paper (S.S. #903) for routine screening is used without autoclaving or any other treatment. Blood standards, each containing a known amount of galactose, 5, 10, 20, and 40 mg/100 ml, are prepared.

Preparation of the Test Plate

The plastic tray for inhibition assay for amino acids is used. If there is any problem of level surface, the bottom of the tray can first be made flat by coating it with a layer of Epon® embedding plastic.*** These trays can be simply washed and reused without sterilization.

The test plate is then prepared as follows: Mix in order: 25.0 ml of base medium, 10 to 15 ml of

1 *N* HCL, 1.0 ml $CaCl_2$ solution ($CaCl_2 \cdot 2H_2O$— 0.3 g/100 ml), 1.0 ml of thiamine solution (0.05 mg/ml), 1.0 ml KH_2PO_4 solution (1.25 g/100 ml) and water to 80 ml. Add 20.0 ml of 7.5% sodium silicate and mix thoroughly. It is important that the pH of the final mixture is between 7.95 and 8.05 (adjust pH with 1 *N* HCl). Deviations in pH will change the solidity of the silicate. Add 6.0 ml of cell suspension, O.D.$_{550}$ = 2.0 and 1.0 ml of phage suspension. Mix gently and pour the entire content into a tray on a level surface. Allow at least 10 min at room temperature for the plate to set. Discs of unknown blood and graded standards 1/8 in. in diameter are placed on the plate 1.5 in. apart. Since the silicate is soft the discs should be carefully set by forceps and not by machine. They are then incubated overnight at 37°C.

Results and interpretation: The size of the zone of inhibition or growth around the unknown is compared with that around the standards. When the level is 6 mg/100 ml or higher, the blood specimen is examined by the fluorescence test for Gal-1-P-uridyltransferase activity. If the fluorescence test shows absence of activity, the result is reported immediately by telephone. On the other hand, when transferase activity is present, a repeat blood specimen is requested to verify the elevation, and galactokinase deficiency should be considered. When the metabolite inhibition assay is positive, an increased valine concentration in the blood should be ruled out by paper chromatography.

Seliwanoff's Resorcinol Test for Fructose[1]

Reagents: The following reagents are freshly prepared and mixed in the proportions indicated:

Resorcinol 0.5% (w/v)	3.5 ml
Concentrated hydrochloric acid	12.0 ml
Water, to	35.0 ml

Method: To 1 ml of urine in a test tube add 5 ml of the mixed reagent. Place the test tube in a boiling water bath for 10 min.

Results and interpretation: The development of a cherry-red color indicates the presence of

---

*300 to 350 ml activated coconut charcoal, 6 to 14 mesh for 1,500 ml silicate solution.

**Dr. Kenneth Paigen.

***A mixture of 31 ml Epon resin 812, dodecenylsuccinic anhydride, and 5 ml 2,4,6-tri(dimethylaminomethyl)-phenol is poured into a plate and cured at 60°C on a level shelf for 48 hr.

fructose. Traces of the color may be found in normal urine.

SCREENING TEST FOR ENZYME DEFECT

This test for galactose-1-phosphate (Gal-1-P) uridyltransferase activity is the only screening test available for an enzyme defect in sugar metabolism. This is a simple fluorescence spot test developed by Beutler and Baluda[4] for the detection of galactosemia (Gal-1-P-uridyltransferase deficiency). It is an enzymatic assay for erythrocyte Gal-1-P-uridyltransferase activity. Whole blood or a disc of filter paper specimen is incubated with galactose-1-phosphate, uridyldiphosphoglucose (UDGP), and nicotinamide adenine dinucleotide phosphate (NADP). When transferase is present glucose-1-phosphate (G-1-P) is formed from UDGP and Gal-1-P, and G-1-P is then converted to glucose-6-phosphate (G-6-P) by phosphoglucomutase. The oxidation of G-6-P by its dehydrogenase results in the reduction of NADP, and fluorescence of reduced NADP (NADPH) under ultraviolet light serves as an indicator of transferase activity.

Reagents and incubation medium: The following reagents are made and the incubation medium is mixed in the proportion indicated.

Reagents	Stock solution	Volume to make 6 ml of reaction mixture
UDGP	9.5×10^{-3} M (6.70 g/l.)	0.2 ml
Gal-1-P	2.7×10^{-3} M (9.56 g/l. of Gal-1-P dipotassium dihydrate)	0.4 ml
NADP	6.6×10^{-3} M (5.05 g/l.)	0.6 ml
Tris-acetate buffer, pH 8.0	0.75 M (Tris base 90.83 g/l. titrate to pH 8.0 with acetic acid)	2.0 ml
Saponin	1% solution (1 g/100 ml)	0.8 ml
EDTA (Disodium ethylenediaminetetraacetic acid)	4.05×10^{-4} M (0.118 g/l.)	2.0 ml

The solutions can be mixed in the above proportion and stored frozen in small aliquots, e.g., 3 ml or its multiples. Alternatively, EDTA may be stored separately and mixed with the other reagents as needed. The volume in individual tubes depends upon the number of tests performed on each occasion. The frozen mixture will remain stable for at least two months.

Method: Blood collected in a heparinized tube or EDTA tube shows no loss of enzyme activity after storage for one week at room temperature. When blood is collected in an EDTA tube, EDTA should be omitted from the incubation mixture since a concentration of EDTA greater than 0.1 mg/ml will inhibit the reaction. Whole blood impregnated on filter paper can be used as well. Transferase in blood filter paper specimens

*Linbro Chemical Company

remains active for at least ten days at room temperature.

Incubation is carried out in a disposable test tube or in a disposable plastic tray containing 72 or 90 individual wells,* each well having the capacity of 0.4 ml.[5] The tray is preferable if a large number of specimens are to be tested. Either 10 μl of whole blood or a 3/8 in. disc of filter paper blood specimen is used. One tenth milliliter of the incubation medium (10 vol of blood) is pipetted into the test tube or the well. The mixture is then incubated at $37°$ for 2 hr when whole blood is used, or for 3 hr when filter paper discs are used. At zero time and at the end of the incubation, a small amount of reaction mixture is transferred to Whatman No. 1 filter paper by dipping one end of a drinking straw into the reaction mixture and then spotting this solution on the filter paper. The spots are allowed to dry before examination under long-wave ultraviolet light.

Hemoglobin in the incubation mixture has a quenching effect on fluorescence, making it difficult to distinguish a much reduced fluorescence from an absence of it. Scherz et al.[6] modified the spotting technique by dipping an ammonium sulfate-treated filter paper tab into the incubation mixture. The hemoglobin and other proteins are precipitated with ammonium sulfate at the end of the filter paper tabs and the solution containing NADPH is drawn a few millimeters higher. This permits clear visualization of the NADPH fluorescence without quenching by hemoglobin. On 1 piece of Whatman No. 1 filter paper there are 12 tabs that are properly spaced according to the measurements of the incubation wells in the tray. These tabs are dipped into 12 wells simultaneously.

Results and interpretation: Lack of fluorescence indicates that transferase activity is less than 25% normal value (Figure 3-1). A homozygote for Duarte variant has 50% of the normal transferase activity. A heterozygote for both galactosemia and Duarte variant has only 25% of the normal transferase activity. Such heterozygotes who are clinically asymptomatic may show no fluorescence by the spot test and may be erroneously diagnosed as galactosemic patients. When incubation is extended to 5 hr or more, the blood spot with 25% transferase will fluoresce faintly, whereas that from a true galactosemic patient will not. However, even with the improved spotting technique,[6] differentiation is not always possible. In such cases, quantitative measurement of the transferase activity is necessary. Testing for galactose in urine is a simple and effective way to rule out the diagnosis of galactosemia.

There have been no known false-negative results with this test. False-positive results (i.e., lack of fluorescence) can occur when the blood filter paper specimen is exposed to the relatively high environmental temperature and high humidity in the course of mailing.[5] The enzyme is partially or completely destroyed under these circumstances. Occasionally, the reason for a false-positive result is unclear.

Since this is an enzyme assay not dependent upon the accumulation of galactose, this test can be performed on cord blood.

Heterozygotes for galactosemia or homozygotes

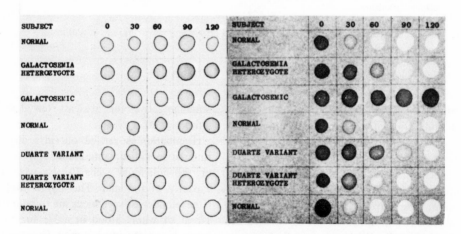

FIGURE 3-1. Fluorescence test for galactose-1-phosphate uridyltransferase deficiency (galactosemia). The photograph at the left shows spotted filter paper as it appears under white light. At the right, the same filter paper has been photographed under long-wave ultraviolet light illumination. The numbers represent time in minutes. At 60 min, the fluorescence of blood samples from a heterozygote for galactosemia and from a homozygote for Duarte variant is considerably less than normal or a heterozygote for Duarte variant. At 120 min, there is still no fluorescence in the blood sample from a galactosemic patient, whereas all other blood samples result in good fluorescence. (Reproduced from Beutler, E. and Baluda, M. C., A simple spot test for galactosemia, *J. Lab. Clin. Med.*, 68, 137, 1966, with permission from the authors and the publisher.)

for Duarte variant may be detected by spotting the incubation mixture after 60 min in incubation, at which time fluorescence of these samples is considerably less than normal (Figure 3-1).

CHROMATOGRAPHY

The chromatographic technique is used for further study when a urine specimen gives a positive reaction to the anthrone test or a positive reaction to Clinitest, and when the reducing substance is not glucose as tested by the glucose oxidase method. A paper chromatographic method is described below. With slight modification, the same technique can be used on thin-layer chromatography.[7],[8]

Quantity of Specimen and Standards
The volume of urine to be spotted is determined by its reaction with Clinitest:

Blue (–)	100 μl
Green (1+ to 2+)	50 μl
Yellow (3+)	25 μl
Orange (4+)	10 μl

A standard solution containing a mixture of sugars is made up in 10% isopropanol at a concentration of 500 mg/100 ml each. Ten microliters of this solution can be spotted and run along with the urine specimens. The relative position of the unknown can be compared with the standards (Figure 3-2).

A solvent system, ethyl acetate-pyridine-water (12:5:4), gives good separation of the clinically important sugars. Since the R_f values of the sugars are low, it is best to run a descending chromatogram overnight (16 to 18 hr) using a 26-in. length paper. The lower end of the paper is serrated to allow even dripping of the solvent.

The BuAc solvent used for amino acid separation has also been used for sugar by Carson and Neely.[10] However, we found the resolution unsatisfactory.

Location Reagents
Aniline-phthalate Reagent[11]
Phthalic acid 1.66% (w/v) in acetone can be prepared and stored at 4°C for weeks. Add aniline 1 ml to 100 ml of the phthalic acid solution before use. The sugar chromatogram is dipped and heated at 100°C for 20 min.

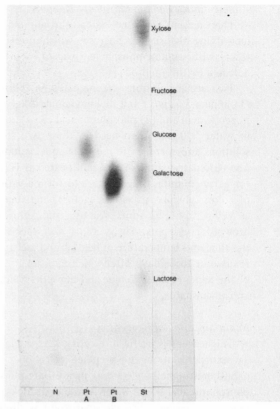

FIGURE 3-2. Descending chromatogram of sugars, developed in ethyl acetate-pyridine-water (12:5:4) overnight, and stained with aniline phosphate reagent. Patient A, Fanconi syndrome; Patient B, galactosemia; N, normal, ST, standard mixture of sugars, 50 μg each. Note the poor color reaction of fructose.

Aniline Phosphate Reagent
The ingredients are mixed in the following order and stored at 4°C: aniline 20 ml, water 200 ml, acetic acid 180 ml, phosphoric acid 10 ml. Before use 2 vol of this reagent are mixed with 3 vol of acetone. The chromatogram is dipped through this reagent and heated for several minutes at 100°C. The optimum time and heating temperature can be determined by trial and error.

These two reagents are used as general location reagents for carbohydrates and give similar results (Figure 3-2). The aniline phosphate reagent is perhaps a little more sensitive. Glucose, galactose, and lactose appear as brown spots, fructose, xylose, and xylulose as reddish-purple spots. The colors are stable for several weeks. Fructose and other ketoses stain poorly and sucrose does not react at all.

Aniline-diphenylamine Phosphate Reagent[8]

This reagent is used as a further aid in identifying the sugars. Sucrose, which does not react with aniline-phosphate reagent, can be detected by this method.

Two stock solutions are prepared as follows: (1) aniline 5% (w/v) and diphenylamine 5% (w/v) in acetic acid and (2) phosphoric acid (85%) 20 ml in water 100 ml. When needed, these two stock solutions and acetone are mixed in equal volumes. The chromatogram is dipped and heated at 100°C for a few minutes. Sugars react to form a variety of colors, glucose, galactose, lactose, and sucrose as yellowish-brown spots, fructose and xylose as brownish-grey spots. These colors may vary with the time and temperature of heating and are stable for two or three days, after which the background of the paper changes to blue and the spots become indistinguishable.

Naphthoresorcinol Reagent

This reagent is useful in locating ketose sugars. Two stock solutions are prepared as follows: (1) naphthoresorcinol 0.2% (w/v) in acetone and (2) phosphoric acid 9% in water. These two stocks are mixed as needed. The chromatogram is dipped through this reagent and heated at 95°C for several minutes in an oven in which the beaker of hot water has been standing for some time. Keto-pentoses are greenish-brown in color, other ketoses (e.g., fructose) are reddish-brown, and uronic acids are blue on a pale pink background. Aldoses do not react unless present in large quantities.

CLINICAL APPLICATION AND INTERPRETATION

With these methods no sugar will be detected in the urine of a normal individual. Neonates, particularly premature infants, may excrete small amounts of glucose in the urine in the first few weeks of life. Woolf[3] has found galactose up to 80 mg/100 ml of urine in newborns two to three weeks of age. Beyond the first month of life the presence of sugar in the urine is usually abnormal, the exception being that lactosuria sometimes occurs in pregnant or lactating women. The ingestion of excessive amounts of nonutilizable sugars leads to the occasional appearance of these sugars in the urine.

In the majority of metabolic disorders of sugar (with the exception of glucose), both the clinical symptoms and laboratory findings of sugar in urine will be detected only when the offensive sugar is ingested, regardless of whether the disorder is due to a metabolic block or a transport defect. For instance, galactosemia is found only after the affected infant has been fed milk which contains lactose, and fructose intolerance most often begins with weaning and the addition of sucrose to the diet.

Probably the most frequent abnormality in screening is glucose in the urine, especially in the adult population in which diabetes mellitus is common. Glycosuria with normal blood glucose indicates a renal transport defect; it may occur alone or in combination with other sugars.

Large amounts of galactose in blood and urine suggest the possibility of a galactose metabolic defect, either Gal-1-P uridyltransferase deficiency or galactokinase deficiency. The diagnosis of Gal-1-P uridyltransferase deficiency, often referred to as classical galactosemia, can be suggested by the clinical symptoms and confirmed by a test for the enzyme. Infants with this disease present with liver disease in early infancy (Table 1-3) and may have the complication of *E. coli* sepsis in the neonatal period.[5] Immediate dietary therapy is essential to the well-being of the infant. Thus, the finding of galactose in blood and/or urine should be reported to the clinician as soon as possible. A test for reducing substances in urine (Clinitest) is a simple and reliable way of detecting galactosuria and should be performed at the doctor's office or on the ward while waiting for results of the confirmatory enzyme test. As a matter of fact, due to the rapid progression of the disease in some infants, it is advocated that a test for the presence of reducing substance in the urine should be a routine procedure for all newborns with jaundice before discharge from the nursery. In the case of a high-risk baby (e.g., the sibling of a known galactosemic patient), the diagnosis should be made as soon as possible by testing the transferase activity in cord blood, and treatment started before any milk feeding.

Whenever galactosemia is found and the transferase deficiency can be ruled out, galacto-kinase deficiency should be considered. Patients with galactokinase deficiency develop cataracts in early infancy, and prompt dietary treatment may prevent this.

Galactose accumulation of acquired origin may

occur in certain types of liver disease in which the ability to metabolize galactose is decreased.

Fructose gives a positive reaction to Clinitest and can be identified by a resorcinol test and by paper chromatography. The clinical and biochemical features in hereditary fructose intolerance (Table 1-3) and the so-called "hereditary tyrosinemia" (Table 1-1) are sometimes similar in many respects. In both, increased tyrosine, methionine, and fructose may be found in urine. The finding of fructosuria should initiate a search for abnormal amino acid patterns. Fructose in the urine may also indicate a benign entity of essential fructosuria which gives no clinical manifestations (Table 1-3).

The presence of both fructose and galactose in the urine, together with a clinical history suggestive of hypoglycemia, leads to the diagnosis of familial galactose and fructose intolerance (Table 1-3). The finding of disaccharides (sucrose and lactose) in the urine may be associated with gastrointestinal symptoms such as vomiting, diarrhea, etc. due to a variety of causes. In such cases, omission of these sugars from the diet may alleviate the symptoms. Infants with intolerance to lactose may also have generalized aminoaciduria that will disappear following a change to a lactose-free diet. Since the amount of lactose in the urine of these infants may not be large enough to be detected by Clinitest, the urine should be chromatographed for sugar if clinical history strongly suggests disaccharide intolerance.[10] Children who are intolerant to all dietary carbohydrates may have the rare glucose-galactose malabsorption syndrome (Table 1-4). These two sugars are found in the stool, and glucose is found in the urine.

Xylulose is a reducing substance which gives a positive reaction with Clinitest. It moves faster than xylose in ethyl acetate-pyridine-water. This sugar is found in patients with pentosuria, a benign asymptomatic disorder found mainly in Jews from Eastern Europe.[12] Much smaller amounts of xylose may also be found in the urine. The presence of xylose could be an artifact of the preparative procedure of chromatography. The significance of the finding of pentosuria is that it may be erroneously diagnosed as diabetes mellitus. On the other hand, these two conditions may coexist. Since Clinitest is nonspecific, the nature of the reducing substance should be established by a more specific test such as the glucose oxidase method or chromatography before any treatment for diabetes mellitus is initiated.

Sialic acid also reduces copper sulfate and gives a positive reaction with Clinitest. This compound has been found in the urine of a patient.[13] Further identification of sialic acids may be obtained by paper chromatography.*[14]

---

*Solvent: butanol:propanol:0.1 N hydrochloric acid (1:2:1); location reagent: dissolve 0.5 g p-dimethylaminobenzaldehyde and 5 g trichloracetic acid in 50% ethanol, and dilute with 60 ml butanol. The paper is dipped or sprayed with this reagent and heated at 100°C for 10 to 15 min. N-acetylsialic acid (or N-acetylneuraminic acid) has an R_f of 0.44 and appears as a purple spot. The color sometimes does not develop for 24 hr.

REFERENCES

1. **Harper, H. A.,** *Review of Physiologic Chemistry,* 12th ed., Lange Medical Publications, Los Altos, Calif., 1969, 11.
2. **Berry, H. K., Leonard, C., Peters, H., Granger, M., and Chunekamrai, N.,** Detection of metabolic disorders: Chromatographic procedures and interpretation of results, *Clin. Chem.,* 14, 1033, 1968.
3. **Woolf, L. I.,** Large-scale screening for metabolic disease in the newborn in Great Britain, in *Phenylketonuria and Allied Metabolic Diseases. Proceedings of a Conference Held at Washington, D. C., April 6–8, 1966,* Anderson, J. A. and Swaiman, K. F., Eds., U.S. Department of Health, Education, and Welfare, Washington, D.C., 1967, 50.
4. **Beutler, E. and Baluda, M. C.,** A simple spot test for galactosemia, *J. Lab. Clin. Med.,* 68, 137, 1966.
5. **Shih, V. E., Levy, H. L., Karolkewicz, V., Houghton, S., Efron, M. L., Isselbacher, K., Beutler, E., and MacCready, R. A.,** Galactosemia screening of newborns in Massachusetts, *N. Engl. J. Med.,* 284, 753, 1971.
6. **Scherz, R., Pflugshaupt, R., and Bütler, R.,** Improved method of mass screening for galactosemia, *Clin. Chim. Acta,* 39, 109, 1972.
7. **Szustkiewicz, C. and Demetriou, J.,** Detection of some clinically important carbohydrates in plasma and urine by means of thin-layer chromatography, *Clin. Chim. Acta,* 32, 355, 1971.
8. **Raadsveld, C. W. and Klomp, H.,** Thin-layer chromatographic analysis of sugar mixtures, *J. Chromatogr.,* 57, 99, 1971.
9. **Menzies, I. S. and Seakins, J. W. T.,** Sugars, in *Chromatographic and Electrophoretic Techniques, Vol. 1, Chromatography,* 3rd ed., Smith, I., Ed., John Wiley & Sons, New York, 1968, 310.
10. **Carson, N. A. J. and Neely, R. A.,** Disaccharide intolerance in infancy, *Arch. Dis. Child.,* 38, 574, 1963.
11. **Efron, M. L., Young, D., Moser, H. W., and MacCready, R. A.,** A simple chromatographic screening test for the detection of disorders of amino acid metabolism, *N. Engl. J. Med.,* 270, 1378, 1964.
12. **Hiatt, H. H.,** Pentosuria, in *The Metabolic Basis of Inherited Disease,* 3rd ed., Stanbury, J. B., Wyngaarden, J. B., and Fredrickson, D. S., Eds., McGraw-Hill, New York, 1972, 119.
13. **Montreuil, J., Biserte, G., Strecker, G., Spik, G., Fontaine, G., and Farriaux, J. P.,** Description d'un nouveau type de méliturie: la sialurie, *C. R. Acad. Sci., Paris,* 265, 97, 1967.
14. **Svennerholm, E. and Svennerholm, L.,** Quantitative paper partition chromatography of sialic acids, *Nature,* 181, 1154, 1958.

Chapter 4

ORGANIC ACIDS

INTRODUCTION

Organic acids have gained clinical interest during the past decade with the discovery of genetic disorders affecting primarily the metabolism of organic acids. General techniques for studying these organic acids that are simple and suitable for routine clinical use have been developed only recently. Several of these currently in use for screening are described here.

CHEMICAL TESTS

Tests for Methylmalonic Acid (MMA)

Giorgio and Plaut[1] described a colorimetric method for quantitative measurement of urinary MMA in 1965; this method was later modified to a simple spot test for screening.[2]

Screening Test

Reagents: *p*-Nitroaniline, 0.1% (w/v) in 0.16 *N* HCl. Stable for six months when stored in a dark glass bottle.

Sodium nitrite, 0.5% (w/v). Stable for at least two months when stored at 4°C.

Sodium acetate buffer, 1 *M*, pH 4.3. Dissolve sodium acetate trihydrate 13.6 g in distilled water and bring the final volume to 100 ml. Add to it 158 ml of 1 *M* acetic acid. Check the pH and adjust if necessary to 4.3 ± 0.02. Stable for at least two to three months when stored at 4°C.

Sodium hydroxide, 8 *N*. Dissolve 32.0 g sodium hydroxide in distilled water and bring the final volume to 100 ml.

Standard solution: Methylmalonic acid 0.025 *M* solution is prepared by dissolving 147.5 mg in 50 ml of distilled water. Add 1 drop of 6 *N* hydrochloric acid to increase the stability of the solution. Stable for at least six months when stored at 4°C.

Method: To 1 drop of urine in a test tube, add 15 drops of 0.1 *N p*-nitroaniline solution, and then 5 drops of 0.5% sodium nitrite solution. The tube is briefly shaken and *p*-nitroaniline is partially or completely decolorized. Twenty drops of the sodium acetate buffer, pH 4.3 are added. The pH

of the mixture should be between 3.8 and 4.3 for optimal color development. (It is necessary to test the pH only when a new batch of reagent is used.) The tube is immediately placed in a boiling water bath for at least 1 min. Periods up to 2 to 3 min will have no appreciable effect on color development. The solution is alkalinized immediately by the addition of 5 drops of 8 *N* sodium hydroxide. Occasionally more sodium hydroxide, up to double the amount, is needed. The color reaction is noted immediately. With each series of urine assays, a reagent blank (1 drop of water instead of urine) and a methylmalonic acid standard should be included for comparison.

For urine specimens impregnated on filter paper (S.S. #903) or PKU testing paper, two discs of 3/8 in. diameter (approximately 25 *μ*l each) are used. After the addition of *p*-nitroaniline solution, the tube is shaken by hand for about 30 sec before adding acetate buffer. The remainder of the procedure is the same as for liquid urine.

Results: An emerald green color developed immediately upon alkalinization indicates a positive reaction. The color is stable for at least 10 or 15 min. Methylmalonic acid present at 60 mg/100 ml urine is detectable. A urine specimen containing no methylmalonic acid gives a brownish color. When a negative result is obtained, additional drops of 8 *N* sodium hydroxide may be added to insure that the pH of the solution is over 12.5. Malonic and ethylmalonic acids give a similar but weaker green color at the same concentration as methylmalonic acid. Some drugs react with the diazo reagent to give a brownish color, thus decreasing the sensitivity of the test. If an equivocal result is obtained, another urine specimen should be collected at least three days after the patient has been taken off medication.

MMA is not detectable in normal urine by this screening test, and its presence is abnormal. Further discussion of methylmalonic aciduria of the genetic and acquired forms will be found in a later section.

Quantitative Measurement

A positive MMA screening test should be

followed by a quantitative measurement.

Reagents: *p*-Nitroaniline, 0.075% (w/v) in 0.2 *N* hydrochloric acid.

Sodium nitrite, 0.5% (w/v)

Acetate buffer, 1 *M* and 0.2 *M*, pH 4.3

Diazo reagent: Add 4 ml of sodium nitrite solution to 15 ml *p*-nitroaniline solution. The mixture is cooled on ice and 4 ml of 0.2 *M* acetate buffer are then added. This reagent should be kept at 4°C and used the same day.

Method: Urine is collected without preservatives and kept frozen until analysis. Adjust the pH to 6.5 and centrifuge if necessary to remove any precipitate. Apply 5 ml of urine to a Dowex® 3-X4 column (chloride form) 1 x 3 cm, wash twice with 50 ml distilled water (5 to 15 ml/min), and elute MMA by gravity with 20 ml of 0.1 *N* hydrochloric acid. Prepare two large test tubes, one containing 2 ml of eluate and one containing 1 ml of eluate plus 1 ml of 0.1 *N* hydrochloric acid (the latter tube may be omitted if the qualitative test of the urine is only weakly positive). Add 3 ml of 1 *M* acetate buffer, pH 4.3, and mix; then add 3 ml of cold diazo reagent and bring the pH to 4.0 for optimal color development. Heat the tubes in a water bath at 94°C for 3 min, and add 2 ml of 3.0 *N* sodium hydroxide. The tubes are then removed from the water bath, stoppered, mixed, and cooled at room temperature for 10 min. A reagent blank containing 2.0 ml of 0.1 *N* hydrochloric acid instead of the column eluate of the urine, and a standard containing 2.0 ml of 0.1 *N* hydrochloric acid and 0.02 ml of 0.05 *M* MMA are handled in the same manner. The optical absorbance is read at 620 nm within 2 hr. The optical density is linear between 0.05 and 1 μmol of MMA. When 1 ml of the column eluate is used, MMA up to 4 μmol/l. (approximately 600 mg/l.) can be measured accurately.

Results and interpretation: By this method MMA is found at a concentration of 5 ± 3.6 mg/l. in normal urine.[1] A number of compounds have been tested for possible interference with the colorimetry;[1] among those that develop a significant color with the diazo reagent, acetoacetate, urate, creatinine, and ascorbate are not eluted from the column in the same fraction and present no problem. Malonate, the only compound that interferes appreciably, is not usually found in the urine. Among those which do not interfere are propionic acid and bilirubin.

This colorimetric method for the measurement

of MMA is not as sensitive as the GLC method. The latter is preferred for determining mild increases.

a-Keto Acids

The presence of excessive keto acids in urine can be demonstrated by the 2,4-dinitrophenylhydrazine test described in Chapter 3. These dinitrophenylhydrazones can be identified by chromatography and measured quantitatively as described in later sections.

A few tests suitable for the specific qualitative and quantitative determinations of pyruvic acid, oxaloacetic acid, a-ketoglutaric acid, and acetoacetic acid are given in a review by Neish.[3] Measurements of pyruvic acid and lactic acid using a spectrophotometric[4] or fluorometric[5] method are simple enough that they are now routine procedures at clinical chemistry laboratories.

Hydroxyphenolic Acids

These compounds are derivatives of phenylalanine and tyrosine and are increased in PKU (*o*-hydroxyphenolic acids) and in hypertyrosinemia and tyrosinosis (*p*-hydroxyphenolic acids).

Ferric chloride test (see Chapter 3).

Nitrosonaphthol test. This is a qualitative test for tyrosine and its *p*-hydroxyphenolic derivatives (see Chapter 3).

Modified Millon test. This test measures tyrosine and its *p*-hydroxyphenolic derivatives.

Qualitative test:[6] Millon reagent is prepared by dissolving 10 g mercury in 11 ml of concentrated nitric acid and then diluting with 22 ml of water. Mix 2 drops of urine with 2 drops of the reagent in a white porcelain dish. The development of pink or pinkish-brown color is a positive reaction.

Quantitative test:[7-9]

Reagents: Sulfuric acid, 3 *M* and 3.5 *M*

Mercuric sulfate, 15% (w/v) in 3 *M* sulfuric acid

Sodium nitrite, 0.75 *M*

Method: Acidify 1 ml urine with 1 ml of 3 *M* sulfuric acid and extract the phenolic acids with 10 ml (10 vol) ethyl ether (or ethyl acetate). Transfer 5 ml of the ether phase and mix with 5 ml water. Remove ether completely with a stream of air or nitrogen and make up volume to 5 ml with water. Transfer 2.5 ml of the extract in a test tube, add 0.2 ml of 3.5 M sulfuric acid, and then 0.5 ml of mercuric sulfate. Heat in a boiling water bath for 8 min, and cool in running cold water. Transfer contents to a cuvette and read absorbance at 500 nm as blank. Then add 20 μl sodium nitrite reagent and after 4 min read absorbance again at 500 nm. Standards of p-hydroxyphenylpyruvic acid (pHPPA) 0, 0.25, 0.50, and 1.0 μmol/ml are prepared. Molar equivalents of tyrosine, pHPPA, p-hydroxyphenyllactic acid (pHPLA), and p-hydroxyphenylacetic acid (pHPAA) give similar red colors of about equal intensity; therefore, any one of these compounds can be used as a standard. For preparation of a standard curve, 1 ml each of the standard solution is handled in the same manner as the urine, and its optic absorbance is plotted.

When the total amounts of urinary tyrosine and its metabolites are to be measured, the step of ether extraction can be omitted and 0.5 ml urine used directly. The volume is made up to 2.5 ml with water before the addition of sulfuric acid and mercuric sulfate.

Results and interpretation: In normal urine only a small amount of tyrosine is present. Both hypertyrosinuria and hypertyrosyluria are found in hereditary tyrosinemia (or tyrosinosis), transient neonatal tyrosinemia, and ascorbic acid deficiency. Urine from patients with generalized hyperaminoaciduria gives a positive Millon reaction as a result of increased excretion of tyrosine. Derivatives of o-hydroxyphenolic acids are not detectable by this method.

PAPER CHROMATOGRAPHY

In general, either acidic or basic solvents give more compact and well-defined spots of organic acids than do neutral solvents. In laboratories where paper chromatography has been set up in routine use for other compounds such as amino acids, it can be used as a screening procedure for a number of abnormal organic acids.

Apparatus
The same apparatus are used as for amino acid separation listed in Chapter 2.

Preparation of Specimens
For screening purposes urine samples can be applied directly. Serum is not suitable for screening because the abnormalities are relatively mild and not easily detectable. Once an abnormality is found by screening, the organic acids can be isolated for further study. The keto acids, except imidazole and phenyl-pyruvic acids, are unstable compounds and should be converted to their dinitrophenyl-hydrazones (DNP-hydrazones) before chromatography.

Extraction[10]
To 5 ml of urine acidified with 0.5 ml of concentrated hydrochloric acid in a 50-ml conical centrifuge tube with a screw cap, add 25 ml of ethyl acetate or ether and 10 g of anhydrous sodium sulfate. Shake the tube vigorously for a few minutes and then leave it at 4°C for 1 hr. Twenty milliliters of ethyl acetate layer are transferred from the tube containing the cake of sodium sulfate and are evaporated to dryness in a rotary evaporator at 40°C.* Dissolve the residue in 0.4 ml of 50% isopropanol (0.1 ml = 1 ml urine). The extract contains all ether-soluble organic acids and neutral phenols.

Column Chromatography[11]
A column 4 x 40 mm of Dowex 2-X8, 60-80 mesh in the formate form is prepared by pouring 40 ml of 1 M sodium formate followed by 200 ml of water. Five milliliters of urine are passed slowly down the column followed by 15 ml of water. The organic acids are eluted with 10 ml of 12 N formic acid and the eluate is concentrated to 0.5 ml by placing the flask in a closed jar containing a mixture of anhydrous calcium chloride and sodium hydroxide (2:1).

a-Keto Acid Hydrazones
The hydrazones of a-keto acids can be prepared by reacting a-keto acids with 2,4-dinitrophenyl-hydrazine.

Two milliliters of 0.1% 2,4-dinitrophenyl-hydrazine (DNPH) in 2 N hydrochloric acid are added to 2 ml of urine. Mix and then allow the reaction to proceed for 30 min at room temperature (20 to 25°C). Extract with 4 ml of peroxide-free diethyl ether** twice and combine the extracts which include the unreacted DNPH, neutral hydrazones, and acidic hydrazones. Extract the acidic hydrazones from the organic

---

*If ether is used it can be evaporated by placing the tube in a warm water bath.

**The hydrazones are better extracted by ether than by ethyl acetate.

solvent layer with 8 ml of 10% sodium carbonate three times, or repeat the step until the sodium carbonate solution becomes colorless. The sodium carbonate extract is carefully acidified by the dropwise addition of cold concentrated hydrochloric acid. Extract with an equal volume of ether. The ether extract is dried over anhydrous sodium sulfate and evaporated to dryness in vacuo or in a stream of air. Dissolve residue in 0.5 ml of ethanol-ethyl acetate (1:1).

Solvents

Butanol-acetic acid-water (12:3:5) — This solvent can be used as a preliminary screening procedure; in it most of the organic acids have high R_f values.

Ethanol-ammonia (25% ammonium hydroxide)-water (100:16:12) (EtAm) — This basic solvent, originally described for methylmalonic acid,[12] can be used for screening.

Isopentyl formate-formic acid (220:40) (IpfF) — This solvent with just enough water added to obtain a persistent turbidity[13] gives a good separation of nonaromatic acids.

n-Butanol-ethanol-water (BuEt) (70:10:20)[14] — This solvent is similar to the upper phase of the mixture used by Cavallini and Frontali.[15] If it becomes cloudy in cold water, add a few extra drops of ethanol. This solvent is useful in the separation of DNP-hydrazones of keto acids.

Isopropanol-ammonium hydroxide-water (IPrAm) (120:15:15)[6] — This solvent is useful in the separation of phenolic acids.

Benzene-acetic acid-water (125:72:3) (BzA)[10] — This solvent should be used in a tightly sealed tank to prevent rapid evaporation of benzene. Preferably the paper is placed in a dry tray in a closed tank, and the solvent is added by means of a tube inserted through a small hole in the top plate. This solvent is temperature-sensitive, and the R_f values may vary with each run. Therefore, standards should be run in parallel. Propionic acid may be used in place of acetic acid in the same proportions as originally described by Armstrong et al.[16] The disadvantage of propionic acid is its unpleasant odor.

Solvents suitable for separation of indoles and p-hydroxyphenolic acids have been described in Chapter 2. The HVE and BuAc system is quite adequate for imidazoles.

Location Reagents

Bromcresol green reagent — Bromcresol green is prepared as a 0.5% (w/v) solution in 95% ethanol. Titrate with 1 N sodium hydroxide until the color is blue-green. Mix 1 vol of this solution with 4 vol of acetone before use.

The dried chromatogram is rapidly drawn through the reagent and laid flat on a clean sheet of paper. Acidic compounds appear as yellow spots, while basic compounds appear as blue spots on a green background. This pH indicator can be used as a general location reagent for screening. Since the color may change, the spots should be marked. Although it has been suggested that other reagents can be applied over bromcresol green, we have found it unsatisfactory.

Aniline-xylose reagent[11] — Dissolve 1 g xylose in 3 ml of water, add 1 ml aniline, and make up the volume to 100 ml with methanol.

The chromatogram is dipped into this reagent and hung up in a fume hood for 5 to 10 min to allow methanol to evaporate before the paper is heated at 105 to 110°C for 5 to 10 min. Permanent brown spots appear on a light yellow background. This reagent is a good general method for the location of acidic spots and is useful in screening.

Altman reagent — p-Dimethylamino-benzaldehyde 5% (w/v) in acetic anhydride.[10] The above stock solution is mixed with 4 vol of acetone as needed. The staining procedure should be performed in a fume hood. The chromatogram is rapidly dipped through this reagent and hung up in a hood to dry. Colors begin to appear within 20 min, and the chromatogram should be examined again the next morning. This reagent is specific for aroyl-glycines such as hippuric acid which yield orange or orange-red color and yellow fluorescence under ultraviolet light. Isovalerylglycine and MMA stain yellow with it. Many of the acids in the citric cycle react to form various shades of color. For instance, citric acid and a-ketoglutaric acid are pink in color. Both Ehrlich and Pauly reagent can be applied over this.

Nitraniline reagent — Three stock solutions are prepared as follows: p-nitraniline 1.5 g in 45 ml of concentrated hydrochloric acid and 950 ml of water; sodium nitrite, 5% (w/v); and sodium carbonate anhydrous, 10% (w/v).

Just before use add 0.2 vol of nitrite solution to 10 vol of nitraniline solution. Mix, and add the carbonate solution. This reagent is similar to Pauly reagent but is more sensitive in detecting many phenolic acids. It cannot be used over Ehrlich reagent.

Ultraviolet (UV) light – Chromatograms should be examined before staining. Some phenolic acids fluoresce under ultraviolet light. This method may be useful in differentiating exogenous phenolic acids in urine from other organic acids. Keto acid hydrazones absorb in the long-wave ultraviolet region and appear as dark spots. This is a very sensitive method and can detect amounts less than 1 μg.

Alkali reagent – Sodium hydroxide 2% (w/v) in 95% ethanol.

When the chromatogram is dipped through the reagent, colors of the DNP-hydrazones appear immediately. The shade and intensity of the color vary with the solvent used and fade slowly.

Procedure and Results

For screening purposes, urine equivalent to 30μg creatinine or a 3/8 in. disc of filter paper urine specimen is applied directly without extraction to Whatman 3MM paper, and a one-dimensional ascending chromatogram is developed in either BuAc or EtAm. Spots can be visualized by staining the chromatogram with the bromcresol green reagent or the aniline xylose reagent. For two-dimensional separation, urine extracts are applied to 10-in. squares of Whatman No. 1 or 3MM paper mounted on a Universal frame. A combination of EtAm and IpfF has been used for aliphatic acids such as MMA. For phenolic acids, IprAm followed by BzA is a suitable system.

Normal urine contains many organic acids which can only be visualized on paper chromatograms when extracts of urine containing at least 2 mg of creatinine are applied. Most of the acids in the citric acid cycle are present in small amounts. Only three keto acids, i.e., pyruvic, a-ketoglutaric, and acetoacetic acids, have been found in the urine. The majority of the phenolic acids in normal urine are of dietary or exogenous origin. Fruits, vegetables, tea, and coffee all contribute to the urinary phenolic acids. A significant proportion is produced by intestinal bacteria. The m-hydroxy- and 3-methoxy-4-hydroxy-phenolic acids are all of exogenous origin, and their excretion falls significantly with bowel sterilization.[17] In contrast, the ortho- and parahydroxyphenolic acids are not significantly affected by dietary intake or sterilization of the gut. Con-sequently, caution should be exercised in studying a patient who excretes an unknown or an unusual amount of phenolic acid, and exogenous origin should first be ruled out.

With the small amount of urine used in screening, there is only one large acidic spot, the identity of which is not clear (Figures 4-1 and 4-2). Many organic acids in normal urine are in amounts small enough to be undetectable. Any deviation from the normal pattern is worth further investigation. Urine specimens from patients with known methylmalonic aciduria and isovaleric acidemia were impregnated on filter papers, and a 3/8 in. disc was used for chromatography as described. Both MMA and isovalerylglycine spots were readily visible with the bromcresol green reagent, indicating that urine collected on filter paper is adequate for screening for at least these two disorders.

THIN-LAYER CHROMATOGRAPHY

Thin-layer chromatography (TLC) requires less time than paper chromatography. However, organic acids in urine must be extracted before application. The procedure of extraction is the same as described in the previous section. The absolute R_f values of the organic acids on TLC are not consistent, but the relative mobility remains the same.

Methylmalonic Acid

Dreyfus and Dubé[12] described a TLC method for the detection of MMA. MMA is extracted from acidified urines and is applied to TLC plates coated with silica gel. A one-dimensional ascending chromatogram is developed in the ethanol-ammonia-water solvent, as described in the previous section. The tank is lined with chromatography paper, and development is completed within 2 to 3 hr. The plate is air-dried thoroughly and sprayed with bromcresol green 0.04% (w/v) in ethanol, titrated to blue color with 0.1 N sodium hydroxide. MMA and other organic acids appear as yellow spots.

For quantitative measurement of MMA, the spot is scraped from the plate and transferred to a test tube. Add 1.0 ml of 0.1 N hydrochloric acid to elute MMA, which is then measured by the color reaction described by Giorgio and Plaut.[1] After the final step, the tube is centrifuged to

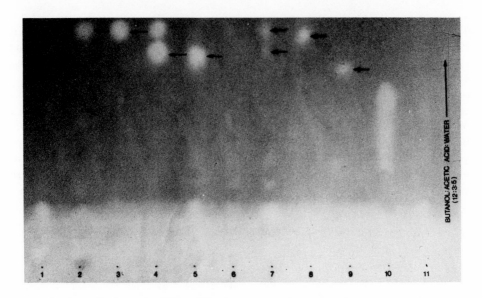

FIGURE 4-1. Paper chromatogram of organic acids developed overnight in butanol-acetic acid-water (12:3:5) and stained with bromcresol green reagent. 1. Normal urine 2. Urine from a patient with isovaleric acidemia; arrow indicates the spot of isovalerylglycine 3. Isovaleryl-glycine standard 4. Methylmalonic acid standard; the top spot is an impurity in the chemical, probably methylsuccinic acid 5. Urine from a patient with methylmalonic aciduria 6. Normal urine 7. Urine from a patient with phenylketonuria; arrows indicate the phenolic acid metabolites 8. Hippuric acid standard 9. Lactic acid standard 10. Pyruvic acid standard (streaky) 11. Normal urine.

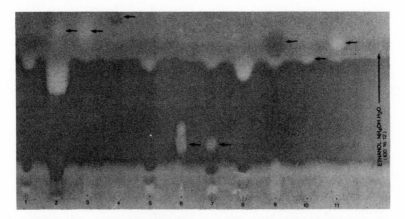

FIGURE 4-2. Paper chromatogram of organic acids developed overnight in ethanol-ammonium hydroxide-water (100:16:12) and stained with bromcresol green reagent. 1. Normal urine 2. Urine from a patient with isovaleric acidemia; isovalerylglycine but not isovaleric acid is visible 3. Isovalerylglycine standard 4. Isovaleric acid standard 5. Normal urine 6. Methylmalonic acid standard; the top spot is probably methylsuccinic acid, an impurity in the chemical 7. Urine from a patient with methylmalonic acidemia 8. Normal urine 9. Propionic acid standard 10. Lactic acid standard 11. Hippuric acid standard.

separate the silica gel, and the supernatant fluid is used for reading.

Isovalerylglycine (IVG)

Ando and Nyhan[18] described a TLC method

for the detection of IVG as a screening procedure for isovaleric acidemia.

An aliquot of a 24-hr urine sample equivalent to 0.7 mg creatinine is placed in a 15 or 20 ml centrifuge tube. Add 0.2 ml of 5 M sulfuric acid for each milliliter of urine. IVG is extracted with 5 ml of a mixture of n-butanol and chloroform (1:5). Centrifuge to separate the two layers if necessary. Two ml of the lower phase are evaporated in a stream of air, and the residue is dissolved in 0.2 ml of the extraction mixture for chromatography. Apply 10 μl of this solution to a TLC plate coated with silica gel, and develop in benzene-isoamyl alcohol-formic acid (70:25:5). The plate is then dried at 100°C for 1 hr and sprayed with bromcresol purple 0.1% (w/v) in ethanol which has been titrated with dilute ammonium hydroxide to a purple color. IVG appears as a yellow spot on a blue-purple background. IVG has an R_f of 0.49 and is separated for hippuric acid with an Rf of 0.60 (Figure 4-3). There is no detectable amount of organic acid in normal urine by this method.

Keto Acid DNP-hydrazones

Dancis et al.[19] described a method for

FIGURE 4-3. Thin-layer chromatogram of isovaleryl-glycine, developed in benzene-isoamyl alcohol-formic acid (70:25:5). Samples from left: hippuric acid standard; isovalerylglycine standard; hippuric acid and isovaleryl-glycine; urine from a patient with isovaleric acidemia; sodium hippurate standard. (Reproduced from Ando, T. and Nyhan, W. L., A simple screening method for detecting isovalerylglycine in urine of patients with isovaleric acidemia, *Clin. Chem.*, 16, 420, 1970, with permission from the authors and the publisher.)

separating primarily the branched chain keto acids found in maple syrup urine disease. The DNP-hydrazones of the keto acids in urine are prepared by the procedure described previously, and are applied to the silica gel-coated TLC plates that have been activated at 110°C overnight. These plates are developed in isomylalcohol-0.25 N ammonium hydroxide (20:1) for 4 to 6 hr. The DNP-hydrazones are yellow in color; no location reagent is needed. Several keto acids give rise to isomeric 2,4-dinitrophenylhydrazones and appear as two spots. The R_f values of the DNP-hydrazones are a-ketoisocaproic acid (2 isomers) 48 and 22, a-keto-β-methylvaleric acid 48, a-ketoisovaleric acid 42, phenylpyruvic acid 46 and 22, and a-ketoglutaric acid 0.

Phenolic Acids

Sankoff and Sourkes[20] described a method for the measurement of homovanillic acid; this method can also be used to study other phenolic metabolites in hereditary tyrosinemia. The phenolic acids are extracted with ethyl acetate and the residue dissolved in methanol. Extracts equivalent to 250 μl of urine are applied to silica gel plates and the plates are developed in the upper phase of benzene-acetic acid-water (2:3:1)* for 80 to 120 min. The plates are air-dried and then sprayed with 1 N solution of Folin's phenol reagent,** followed by 10% sodium carbonate. Homovanillic acid has an R_f of 0.67, pHPPA 0.33, pHPAA 0.56, DOPA 0, and DOPAmine 0.

GAS-LIQUID CHROMATOGRAPHY

Short Chain Fatty Acids

Improved methods for the analysis of short chain fatty acids (SFA, C_2-C_6) by gas-liquid chromatography (GLC) have been developed so that it is now practical to perform such analysis on a large number of specimens for clinical purposes. Recently, Perry et al.[21] and Kurtz et al.[22] described similar techniques for the analysis of SFA in physiologic fluids that eliminate the step of preliminary extraction and significantly reduce the time involved in the preparation of samples. Levy et al.[23] further improved the separation by slight modification of the method, which is described below.

Preparation of Specimen

Serum or plasma (3 to 5 ml) is mixed with an

*Similar to BzA described earlier.
**Fisher Scientific Company

equal amount of distilled water in a 250 ml flat bottom flask. Add 2-methylvaleric acid as an internal standard at 10 μl/ml of serum. The solution is acidified by 1 to 2 ml of 2 M sulfuric acid and is mixed by gentle swirling. The flask is then placed in a distilling apparatus, and steam is bubbled through the specimen for 10 to 20 min or until 30 ml of distillate are collected in a flask packed in ice. Add 1 drop of phenolphthalein solution to the distillate and alkalinize it with 1 to 2 drops of 1 N sodium hydroxide until the solution turns pink. The distillate is then taken to dryness on a rotary evaporator at 60 to 70°, and the residue is dissolved in 2 ml of distilled water. The solution is transferred to a small conical test tube and dried in a simple evaporator under nitrogen for 30 to 45 min at 50 to 60°. The dried specimen is stored at $-20°C$ in the test tube until analysis. For analysis, the residue is dissolved in 0.1 ml of 25% metaphosphoric acid, and 5 to 10 μl of the solution are injected into the GLC inlet with a Hamilton syringe.

Urine and CSF specimens can be prepared in the same fashion as serum.

Apparatus and Method

A gas-liquid chromatograph with a flame ionization detector system can be used. A 6-ft column is packed with SP-1200/H_3PO_4.* Either helium or nitrogen can be used as a carrier gas with a flow rate of approximately 40 to 50 ml/min. Column temperature is 125°C throughout.

Results

Table 4-1 gives the amounts of SFA in normal serum, cerebrospinal fluid (CSF), and urine. Large amounts of acetic acid are present in all three fluids. The concentrations of propionic acid, isobutyric acid, and isovaleric acid are much less than that of acetic acid, while n-butyric, n-valeric, and n-hexanolic acids are occasionally found in serum (Figure 4-4). Interestingly, unlike amino acids, the amounts of SFA are in the same ranges in CSF and serum.[21]

When the serum sample is acidified by concentrated sulfuric acid[22] the amounts of acetic acid and propionic acid are considerably higher than the values presented in the table; presumably, a significant portion of these acids is present in the bound form and is freed by strong acidity.

Patients with isovaleric acidemia and propionic acidemia have marked elevations of these corresponding acids (Figures 4-5 and 4-6), and the diagnosis can easily be made using only 1 ml serum for extraction. Tiglic acid, which has been found in large quantity in the urine of patients with propionic acidemia,[24] appears immediately after β-methylcrotonic acid.

The presence of crotonic acid in specimens containing an excess of β-hydroxybutyric acid is usually an artifact; it is formed by the chemical dehydration of β-hydroxybutyric acid during steam distillation.[25]

n-Valeric acid has been used as an internal standard; however, since a peak in the area of n-valeric acid may be found in normal serum, it is better to use 2-methylvaleric acid instead.

Methylmalonic Acid (MMA), the Aliphatic and Phenolic Acids

Gibbs et al.[26] described a simple and rapid GLC method for the determination of urinary MMA and other mono- to oligo-carboxylic acids in less than 90 min. It can be used as a routine clinical procedure for screening. The method involves no hazardous chemical such as diazomethane, which is an extremely toxic compound and has a tendency to detonate unexpectedly. Instead, trimethylsilyl derivatives are formed with a simple procedure.

Collection and Preparation of Samples

A timed urine specimen is collected and kept at $-20°C$ until analysis. An aliquot of 2 ml urine is saturated with sodium chloride (about 0.5 g) in a 10 ml glass centrifuge tube with a Teflon®-lined screw cap. Adjust the pH to 11 to 12 with a few drops of 1 N NaOH and add 3 to 4 ml of diethyl ether.** The tube is tightly capped and shaken on a small vortex agitator at high speed for 30 sec. Centrifuge if emulsion forms. The extraction is repeated twice. Neutral compounds that may interfere with analysis are removed by the extraction, and the ether layer (top) is discarded. The aqueous phase is then adjusted to pH 2 ± 0.5 with a few drops of 1 N HCl. Three ether extractions are made as before. The retained ether layers are saved, combined, dried over anhydrous sodium sulfate for 5 min, and concentrated to 1 to 2 ml volume in a dry nitrogen stream. The

*Supelco, Bellefonte, Pa.
**Microanalytical quality.

TABLE 4-1

Concentrations of Short Chain Fatty Acids (SFA) in Blood, Cerebrospinal Fluid (CSF), and Urine

	Serum		CSF	Urine
Number of subjects Reference Acid	22 Levy et al.[23] μmol/l. Mean ± SD	31 Perry et al.[21] μg/ml (μmol/l.)[b] Mean ± SD	21 Perry et al.[21] μg/ml (μmol/l.)[b] Mean ± SD	41 Perry et al.[21] μg/mg creatinine Mean ± SD
Acetic	237.07 ± 87.15	4.65 ± 1.66 (77.4) ± (27.6)	6.99 ± 3.33 (101.4) ± (55.5)	2.63 ± 1.79
Propionic	11.22 ± 3.64	0.32 ± 0.37 (4.3) ± (5)	0.21 ± 0.24 (2.8) ± (3.2)	tr − 0.25[a]
Isobutyric	2.25 ± 1.47	0 − 0.22[a] (0 − 2.5)	0 − 0.32[a] (0 − 3.6)	0 − 0.07[a]
n-Butyric	1.31 ± 1.47	0 − 0.29[a] (0 − 3.3)	0 − 0.25[a] (0 − 2.8)	0 − 0.26[a]
Isovaleric	1.28 ± 0.76	0 − 0.42[a,c] (0 − 4.1)	0 − 0.28[a,c] (0 − 2.7)	0 − 0.22[a,c]
n-Valeric	0 − 0.48[a]	−	−	−
n-Hexanoic	0.30 ± 0.32	0 − 0.32[a] (0 − 2.8)	0 − tr[a]	0 − 0.18[a]

[a] Range.
[b] Recalculated from Perry et al.
[c] May also include a-methylbutyric acid

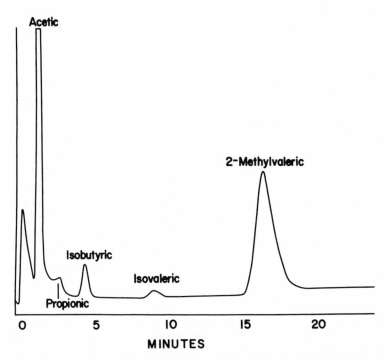

FIGURE 4-4. Short chain fatty acid pattern in normal serum. 2-Methylvaleric acid was added as an internal standard. (Reproduced from Kurtz, D. J. et al., A rapid method for the quantitative analysis of short-chain fatty acids in serum or plasma, *Clin. Chim. Acta*, 34, 463, 1971, with permission of the authors and publisher.)

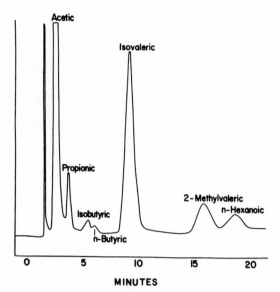

FIGURE 4-5. Short chain fatty acid analysis of serum from a child with isovaleric acidemia. 2-Methylvaleric acid was added as an internal standard. (Reproduced from Kurtz, D. J. et al., A rapid method for the quantitative analysis of short-chain fatty acids in serum or plasma, *Clin. Chim. Acta,* 34, 463, 1971, with permission of the authors and publisher.)

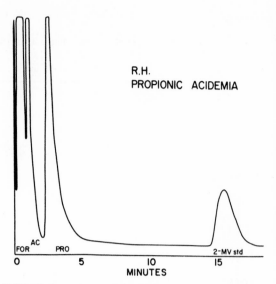

FIGURE 4-6. Short chain fatty acid analysis of serum from a child with propionic acidemia. 2-Methylvaleric acid (2-MV) was added as an internal standard. (Courtesy of Dr. Harvey L. Levy.)

concentrate is then transferred to a silylation tube* and evaporated to near dryness in a nitrogen stream. The derivatizing agent, Tri-Sil/BSA,** approximately 100 μl, is injected through the serum stopper and mixed. An internal standard *o*-hydroxyphenylacetic acid (o-HPAA) 1 to 5 μl of a 1% (w/v) solution in pyridine is also injected through the stopper. The derivatization is carried out by standing the tubes in a 60 to 70° water bath for 15 min. An aliquot of the derivative 0.5 to 2 μl is injected into the GLC inlet by means of a Hamilton syringe.

All glassware should be freed of grease or other organic films by rinsing in 1% warm hydrofluoric acid and then in distilled water.

Preparation of Standards

o-HPAA is chosen as an internal standard because its bis-TMS derivative yields a peak on a portion of the chromatogram where there is no interference with other peaks and because this compound is not normally excreted in any significant quantity except by patients with PKU.

Standards are prepared by mixing MMA and o-HPAA in various ratios, and the mixture is derivatized with Tri-Sil/BSA, approximately 1 mg acid mixture per 1 ml reagent, as described above.

Apparatus and Method

A gas chromatograph with a flame ionization detector system and a single column can be used. A 6-ft column is packed with 3% OV-I on Chromosorb W High Performance 100-120 mesh. The temperature is programmed to start with 80° and to increase by 6°/min to 200°. The detector is set at 290° and the injector at 280°. The carrier gas is helium with a flow rate of approximately 15 ml/min.

*In addition to the commercially available silylation vials with screw caps, Mamer et al.[27] described a simple, inexpensive, and disposable silylation tube that can be made in the laboratory. A short length of Pyrex® tubing is crushed near the center, making one upper and one lower compartment. The upper compartment is used as a reaction chamber and is capped with a serum stopper. The lower half of the tube can be used for support in the water bath and for labeling. Micrograms of samples can be silylated in the reaction chamber made from the tubing of small caliber (such as 3 mm O.D.), and the sample can be easily drawn up by the syringe. The stoppers should be thoroughly washed in acetone to remove extractable potential contaminants.

**Pierce Chemical Company

Results

Figure 4-7 shows the separation of the MMA peak and other organic acids in urine.

A calibration curve is prepared by plotting the ratios of the original concentrations of MMA to o-HPAA in each of the standard mixtures (R_w) as the abscissa, and the ratios of the measured peak heights of MMA to o-HPAA (R_h) as the ordinate. For calculating MMA in the unknown urine, the peaks of MMA and o-HPAA (internal standard) are measured and the ratio (R_h) calculated. R_w can then be found from the calibration curve. The total excretion of MMA is then calculated by this formula:

$$\text{MMA (mg/day)} = \frac{\text{volume of 24-hr urine (ml) x internal standard added } (\mu g) \text{ x } R_w}{100 \text{ x volume of urine extracted}}$$

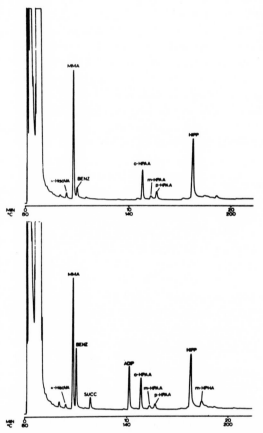

FIGURE 4-7. Gas chromatographic analysis of two urine specimens from a patient with megaloblastic anemia showing the peaks of methylmalonic acid and other organic acids. See text for details of the procedure. o-Hydroxyphenolic acid is added as internal standard. Abbreviations: ADIP, adipic acid; BENZ, benzoic acid; HIPP, hippuric acid; a-HisoVA, a-hydroxyisovaleric acid; HPAA, hydroxyphenylacetic acid; m-HPHA, m-hydroxyphenyl-hydracrylic acid; MMA, methylmalonic acid; SUCC, succinic acid. (Reproduced from Gibbs, B. F. et al., A rapid method for the analysis of urinary methylmalonic acid, *Clin. Chim. Acta,* 38, 447, 1972, with permission of the authors and publisher.)

Other urinary acidic metabolites such as pyroglutamic acid (in pyroglutamic aciduria),[28] pHPAA (in hereditary tyrosinosis), and o-HPPA (in PKU) can also be quantitatively measured by this method. In the latter case, a calibration curve using other internal standards is necessary. More simply, MMA can be added as the internal standard and the same calibration curve can be used. A GLC method described by Wadman et al.[29] is equally applicable in studying the phenolic compounds in abnormal phenylalanine and tyrosine metabolism.

Glycine Conjugates of the SFA

Patients with a metabolic block in the metabolism of SFA who accumulate these acids excrete a significant portion of these SFA as glycine conjugates. Four such conjugates have been identified: isovalerylglycine in isovaleric acidemia,[30] propionylglycine in propionic acidemia,[31] and tiglylglycine and β-methylcrotonylglycine in a patient with β-methylcrotonylglycinuria and β-hydroxyisovaleric aciduria.[32]

An aliquot of urine is acidified to pH 1 with hydrochloric acid and extracted six times with equal amounts of redistilled ethyl acetate. The combined extract is dried over anhydrous sodium sulfate. An internal standard is added, and the extract is evaporated to dryness in a nitrogen stream. The residue is methylated with diazomethane in methanol-diethyl ether (1:10) for GLC analysis.

The method of using three stationary phases as described by Tanaka and Isselbacher[30] is particularly useful for the identification of an unknown peak. A glass column 4 mm by 6 ft with silanized Anakrom,* 80 to 90 mesh, as solid

*Analytical Engineering Laboratories, Hamden, Connecticut

support is used. The three stationary phases and the temperature program for analysis of these acyl-glycines are as follows: 160°C for 5% (w/w) neopentylglycol adipate, 170°C for 5% (w/w) ethyleneglycol adipate, and 190°C for 5% (w/w) SE-30.* The temperature is 220° for both the injector and detector. The column temperature may have to be lowered for resolution of compounds with short retention time, such as propionyl glycine. For routine use, one stationary phase of SE-30 or OV-I is adequate.

Keto Acids

Greer and Williams[33] described a GLC method for quantitative measurement of a-keto and a-hydroxy acids excreted by patients with MSUD. These acids are extracted from a salt-saturated acidified urine specimen and are methylated by reaction with ethereal diazomethane for 15 min. Excessive diazomethane is evaporated under a stream of air until approximately 200 to 400 μl of ethereal solution are left. Methyl benzoate is then added as an internal standard. Several GLC conditions can be used. A 4% QF-1 column is used at 79°C with argon as the carrier gas at a flow rate of 60 ml/min. Three keto acids (a-ketoisocaproic, a-keto-β-methylvaleric, and a-keto-isovaleric) and a-hydroxyisovaleric acid have been measured in the urine of patients. The extraction and analysis can be completed in 2 hr.

GAS CHROMATOGRAPHY – MASS SPECTROMETRY

The combination of gas-chromatography and mass spectrometry (GC-MC) is the best method available for studying an unknown compound. Inasmuch as many steps of the intermediary organic acid metabolism are not well understood, and because of the various possible alternative pathways in the presence of a metabolic block, it can be expected that many unknown compounds in the urine will be found with the increasing use of screening techniques for organic acids. These unknowns can be conveniently identified by this technique using a coupled gas chromatograph-mass spectrometer system. The peaks from GLC are identified by matching the mass spectra of the compounds against known reference spectra. Mamer et al.[27] have obtained data on gas chromatographic separations and characteristic m/e values of authentic heterocyclic, aliphatic, aromatic, and phenolic acids and related compounds that frequently occur in urine or other biological fluids. These mass spectra are tabulated in their report for ready reference. Hutterer et al.[34] recently described the instrumentation and methodology of GC-MS in clinical application; they used a single-step derivative extraction of various compounds including organic acids, amino acids, sugars, drugs, etc. Jellum et al.[35] reported the use of eight GLC-systems in combination with mass spectrometry for the measurement and identification of urinary metabolites including volatile compounds, nonvolatile organic acids, amino acids, peptides, and other hydrolyzable compounds as a technique in the diagnosis and study of metabolic disorders. Three new inborn errors, MMA, β-methylcrotonyl-CoA carboxylase deficiency, and pyroglutamic aciduria, were discovered as a result of such screening in more than 700 patients.

It is conceivable that the combined use of GLC, MS, and computer will be the most valuable screening technique for all metabolic disorders.

PARTITION COLUMN CHROMATOGRAPHY

Organic acids can be eluted from a silica gel column and measured quantitatively by indicator titration using an automatic organic acid analyzer. This technique can be used on a semiroutine basis to study the acidic metabolites in blood and urine.

Apparatus

The analyzer** can be constructed according to the description by Kesner and Muntwyler[36] with the recent modification by Kesner.[37] The basic system consists of a recorder, a photometer, two pumps, a 6-ft mixing coil of 18 gauge Teflon tubing, a narrow glass column 5 x 500 mm, and a five-chambered device for the generation of a gradient solvent system. Simultaneous measurement of ultraviolet absorption and counting of radioactivity can be accomplished with additional modules. The output from the UV detector can be displayed on the same chart of the multipoint recorder.

When a three-channel multipoint recorder,

*Applied Science Laboratories
**May become commercially available in the future.

similar to the one on an amino acid analyzer, is available, analyses of two to three samples can be recorded simultaneously, either one analysis with indicator titration and UV absorption and a second analysis with indicator titration only, or three analyses with indicator titration only.

Indicator

Potassium tetrabromophenolphthalein ethyl ester* is made up at 400 mg/l. in 95% ethanol which has been redistilled over pellets of sodium hydroxide. The indicator solution is delivered at 18 ml/hr and mixed with the effluent from the column in the coil. Acids in the effluent convert the indicator salt to its hydrogen form which has a maximum absorbance at 425 nm.

Column

A glass column** with an end fitting made from a solid Teflon rod is "dry-packed" with 7 g hydrated silical gel by continuous vibration. Air is removed by pumping chloroform through. The

packing procedure takes about 10 min to complete. The silica gel is discarded after each analysis and the packing procedure repeated.

Sample Preparation and Application to Column

Physiologic fluid may be analyzed without preliminary extraction. Up to 200 μl of urine or plasma acidified with 20 μl of 6 N sulfuric acid can be applied directly to the top of the column, or about 300 mg oven-dry silica gel can be added to the sample and the mixture stirred until free flowing occurs. The powder is then added to the top of the column by vibration. Deaerated chloroform is allowed to penetrate the sample by gravity, and a layer of 100 to 200 mg silica gel is added to cover the sample. The column is filled to the socket with chloroform.

Solvent System

A gradient elution system is composed of chloroform and tertiary amyl alcohol. A five-chambered device is prepared as follows:

Chamber	Solvent	Amount
1	Chloroform (C)	90 ml
2	7% t-amyl alcohol (t-AA)/C	Equal weight to chamber 1
3	7% t-AA/C	Equal weight to chamber 1
4	30% t-AA/C	Equal weight to chamber 1
5	50% t-AA/C	Equal weight to chamber 1

The solvent is delivered at 100 ml/hr.

Results and Clinical Application

Kesner[37] has mapped 45 organic acids including those seen in 6 metabolic disorders. These are pyruvic, lactic, propionic, methylmalonic, homogentisic, and orotic acids. In addition, propionylglycine[30] and methylcitrate,[38] both increased in propionic acidemia, have been detected by this technique. Oxaloacetic and oxalosuccinic acids are not stable on these columns. This system is sensitive enough to detect 0.02 to 1 microequivalent of acids; the technique is relatively simple and is specific for acidic metabolites. It is potentially useful as a routine procedure in studying patients with possible organic acid abnormalities.

*Eastman Kodak Company, Rochester, N.Y.
**Fisher Porter Company, Warminster, Pa.

CLINICAL APPLICATION AND INTERPRETATION

Several disorders in the metabolism of organic acids have been known. In the majority of them, recurrent acidosis in early infancy is a prominent feature and may result in early death if undiagnosed and untreated. Therefore, any infant or child with unexplained acidosis should have organic acid screening. Fortunately, the short chain fatty acids and their derivatives have characteristic odors, and bedside diagnosis can often be made by an astute clinician. The diagnosis can be confirmed by laboratory findings. Table 4-2 lists metabolic disorders in which a specific odor is diagnostic. Also included in the table are two disorders not in the SFA metabolism but which also have characteristic odors. The unpleasant odor

TABLE 4-2

Metabolic Disorders Associated with Characteristic Odors

Disorders	Odors
Maple syrup urine disease	Maple syrup
Isovaleric acidemia	Sweaty feet
β-methylcrotonylglycinuria and β-hydroxyisovaleric aciduria	Cat's urine or black currant leaves
Trimethylamine[38]	Fishy odor
Methionine malabsorption syndrome	Oasthouse

of isovaleric acid can be augmented by acidifying the specimens.

Since BuAc solvent is commonly used for amino acids, it is convenient to use the same for organic acid screening. We have also used EtAm as a screening solvent. In general, spots are more compact and better visualized in BuAc than in EtAm, except propionic acid which appears as a well-defined spot when the chromatogram is developed in EtAm but cannot be found in BuAc. The drawback of BuAc is that most of the organic acids have high R_f values. When an abnormal spot is found, the use of different location reagents, separation in another solvent, or measurement by another method is required in identifying the compound. Figures 4-1 and 4-2 show several compounds that may be found in both normal and abnormal conditions.

In BuAc solvent, hippuric acid runs very close to the other organic acids. However, this compound, although a normal constituent of urine, cannot be detected in the amount of urine applied for screening purposes. When in doubt, the chromatogram should be stained with Altman reagent, with which hippuric acid gives an orange color. Lactic acid is better detected in BuAc, whereas pyruvic acid and α-ketoglutaric acid both streak in either BuAc or EtAm. With the rapid advances in GLC technique, it can be predicted that this technique will soon be used as a routine screening procedure for abnormal metabolites. The potential usefulness of the organic acid analyzer is also anticipated.

Isovalerylglycine is excreted in large quantity by patients with isovaleric acidemia, even in the absence of acute clinical symptoms, such as acidosis. This compound can be easily detected on paper chromatograms developed in either BuAc or EtAm, and on TLC it is a better index than isovaleric acid itself. Confirmation of isovaleric acid accumulation in blood and urine as well as its quantitative measurement can be obtained by the GLC method.

MMA is excreted in small quantity by a normal person and can only be detected by quantitative measurement. Its excretion increases tenfold or more in patients with genetic methylmalonic acidurias, i. e., deficiencies of methylmalonyl-CoA isomerase (both B_{12} responsive and unresponsive types) and methylmalonyl-CoA racemase (Table 1-5), and in untreated B_{12} deficiency (pernicious anemia). Urine specimens from two patients with symptomatic methylmalonic aciduria* gave a positive reaction when the chemical screening test was performed with either 1 drop of urine or one 3/8 in. disc of filter paper specimen. One of these patients was a newborn and gave only a weakly positive reaction. In these urines MMA was also detectable by paper chromatography. It is not known how early methylmalonic acid can be detected in the urine of newborn infants before clinical symptoms appear. We have had experience with one infant who died at three months of age with proven methylmalonic aciduria** whose urine specimen collected at one month of age showed no detectable amount of methylmalonic acid by either the chemical or paper chromatographic screening method. A prenatal diagnosis of methylmalonic aciduria has been made by detection of MMA in the maternal urine and amniotic fluid in the third trimester.[40]

Pyroglutamic aciduria is a recently described inborn error in which this compound is excreted in huge quantity. This disorder can probably be detected by paper chromatography and studied further by GLC.[28]

Detection of pyruvic acid and lactic acid by paper chromatography is not satisfactory. It is suggested that these two compounds be determined by chemical methods available at hospital clinical laboratories.

Paper chromatography can be used to detect o-hydroxyphenylacetic and o-hydroxyphenyllactic acids excreted by patients with untreated PKU. Both compounds also react with Pauly reagent. Quantitative measurement of these and other

---

*Courtesy of Dr. Grant Morrow, III, University of Arizona, Tucson, Ariz.

**Dr. Mary G. Ampola, New England Medical Center, Boston, Mass., personal communication.

phenolic acids can be performed by GLC or by silica gel partition chromatography.

In MSUD, three a-keto acids are greatly increased. These are a-ketoisocaproic, a-keto-β-methylvaleric, and a-ketoisovaleric acids. They can easily be detected by a qualitative 2,4-dinitrophenylhydrazine test and can be separated as their DNP-hydrazones by paper chromatography or TLC.

An artifactual abnormal organic aciduria may be found in patients with bacterial infection of the urinary tract.[41] Different strains of bacteria isolated from the urine of patients with urinary tract infection were shown to produce significant amounts of acetic acid, propionic acid, isovaleric acid, etc. Therefore, urine specimens should be frozen immediately after collection. The possibility of bacterial organic aciduria should be considered and should be ruled out before extensive study of a presumed metabolic disorder is undertaken. The finding of an abnormality in the urine but not in the serum suggests contamination.

Several metabolic disorders involving organic acids also involve amino acid abnormalities (Tables 1-1 and 1-5). In addition to the specific amino acid abnormalities, such as increased branched chain amino acids in MSUD and hyperalaninemia in disorders of pyruvate and lactate metabolism, a nonspecific finding of increased glycine in blood and urine has been associated with several organic metabolic defects. Therefore, the diagnosis is often first suspected by amino acid screening and confirmed by finding the abnormal organic acids using one of the techniques described in this chapter. On the other hand, abnormal organic acids may accumulate in the presence of a primary amino acid metabolic defect, and a study of the profile of organic acids in these conditions using GLC or the organic acid analyzer is recommended.

REFERENCES

1. Giorgio, A. J. and Plaut, G. W. E., A method for the colorimetric determination of urinary methylmalonic acid in pernicious anemia, *J. Lab. Clin. Med.,* 66, 667, 1965.

2. Giorgio, A. J. and Luhby, A. L., A rapid screening test for the detection of congenital methylmalonic aciduria in infancy, *Am. J. Clin. Pathol.,* 52, 374, 1969.

3. Neish, W. J. P., a-Keto acid determinations, in *Methods of Biochemical Analysis,* Volume V, Glick, D., Ed., John Wiley & Sons, New York, 1957, 107.

4. Hadjivassiliou, A. G. and Rieder, S. V., The enzymatic assay of pyruvic and lactic acids. A definitive procedure, *Clin. Chim. Acta,* 19, 357, 1968.

5. Olsen, C., An enzymatic fluorimetric micromethod for the determination of acetoacetate, β-hydroxybutyrate, pyruvate and lactate, *Clin. Chim. Acta,* 33, 293, 1971.

6. Berry, H. K., Leonard, C., Peters, H., Granger, M., and Chunekamrai, N., Detection of metabolic disorders. Chromatographic procedures and interpretation of results, *Clin. Chem.,* 14, 1033, 1968.

7. Hsia, D. Y. Y. and Inouye, T., *Inborn Errors of Metabolism, Part 2, Laboratory Methods,* Year Book Medical Publishers, Chicago, 1966, 72.

8. LaDu, B. N. and Zannoni, V. G., The tyrosine oxidation system of liver. II. Oxidation of *p*-hydroxyphenylpyruvic acid to homogentisic acid, *J. Biol. Chem.,* 217, 777, 1955.

9. Bernhart, F. W. and Schneider, R. W., A new test of liver function — the tyrosine tolerance test, *Am. J. Med. Sci.,* 205, 636, 1943.

10. Smith, I., Seakins, J. W. T., and Dayman, J., Phenolic acids, in *Chromatographic and Electrophoretic Techniques, Volume I, Chromatography,* 3rd ed., Smith, I., Ed., John Wiley & Sons, New York, 1969, 364.

11. Nordmann, J. and Nordmann, R., Organic acids, in *Chromatographic and Electrophoretic Techniques, Volume I, Chromatography,* 3rd ed., Smith, I., Ed., John Wiley & Sons, New York, 1969, 342.

12. Dreyfus, P. M. and Dubé, V. E., The rapid detection of methylmalonic acid in urine — a sensitive index of vitamin B_{12} deficiency, *Clin. Chim. Acta,* 15, 525, 1967.

13. Barness, L. A., Young, D., Mellman, W. J., Kahn, S. B., and Williams, W. J., Methylmalonate excretion in a patient with pernicious anemia, *N. Engl. J. Med.,* 268, 144, 1963.

14. Smith, I. and Smith, M. J., Ketoacids, in *Chromatographic and Electrophoretic Techniques, Volume I, Chromatography,* 3rd ed., Smith, I., Ed., John Wiley & Sons, New York, 1969, 330.

15. Cavallini, D. and Frontali, N., Quantitative determination of keto-acids by paper partition chromatography, *Biochim. Biophys. Acta,* 13, 439, 1954.

16. Armstrong, M. D., Shaw, K. N. F., and Wall, P. E., The phenolic acids of human urine. Paper chromatography of phenolic acids, *J. Biol. Chem.,* 218, 293, 1956.

17. Asatoor, A. M., Chamberlain, M. J., Emmerson, B. T., Johnson, J. R., Levi, A. J., and Milne, M. D., Metabolic effects of oral neomycin, *Clin. Sci.,* 33, 111, 1967.

18. Ando, T. and Nyhan, W. L., A simple screening method for detecting isovalerylglycine in urine of patients with isovaleric acidemia, *Clin. Chem.*, 16, 420, 1970.
19. Dancis, J., Hutzler, J., and Levitz, M., Thin-layer chromatography and spectrophotometry of α-ketoacid hydrazones, *Biochim. Biophys. Acta*, 78, 85, 1963.
20. Sankoff, I. and Sourkes, T. L., Determination by thin-layer chromatography of urinary homovanillic acid in normal and disease states, *Can. J. Biochem.*, 41, 1381, 1963.
21. Perry, T. L., Hansen, S., Diamond, S., Bullis, B., Mok, C., and Melancon, S. B., Volatile fatty acids in normal human physiological fluids, *Clin. Chim. Acta*, 29, 369, 1970.
22. Kurtz, D. J., Levy, H. L., Plotkin, W., and Kishimoto, Y., A rapid method for the quantitative analysis of short-chain fatty acids in serum or plasma, *Clin. Chim. Acta*, 34, 463, 1971.
23. Levy, H. L., Erickson, A. M., Lott, I. T., and Kurtz, D. J., Isovaleric acidemia. Results of family study and dietary treatment, submitted for publication.
24. Nyhan, W. L., Ando, T., Rasmussen, K., Wadlington, W., Kilroy, A. W., Cottom, D., and Hull, D., Tiglicaciduria in propionicacidaemia, *Biochem. J.*, 126, 1035, 1972.
25. Gompertz, D., Crotonic acid, an artefact in screening for organic acidaemias, *Clin. Chim. Acta*, 33, 457, 1971.
26. Gibbs, B. F., Itiaba, K., Mamer, O. A., Crawhall, J. C., and Cooper, B. A., A rapid method for the analysis of urinary methylmalonic acid, *Clin. Chim. Acta*, 38, 447, 1972.
27. Mamer, O. A., Crawhall, J. C., and Tjoa, S. S., The identification of urinary acids by coupled gas chromatography—mass spectrometry, *Clin. Chim. Acta*, 32, 171, 1971.
28. Jellum, E., Kluge, T., Börresen, H. C., Stokke, O., and Eldjarn, L., Pyroglutamic aciduria — a new inborn error of metabolism, *Scand. J. Clin. Lab. Invest.*, 26, 327, 1970.
29. Wadman, S. K., Van der Heiden, C., Ketting, D., and Van Sprang, F. J., Abnormal tyrosine and phenylalanine metabolism in patients with tyrosyluria and phenylketonuria; gas-liquid chromatographic analysis of urinary metabolites, *Clin. Chim. Acta*, 34, 277, 1971.
30. Tanaka, K. and Isselbacher, K. J., The isolation and identification of *n*-isovalerylglycine from urine of patients with isovaleric acidemia, *J. Biol. Chem.*, 242, 2966, 1967.
31. Rasmussen, K., Ando, T., Nyhan, W. L., Hull, D., Cottom, D., Donnell, G., Wadlington, W., and Kilroy, A. W., Excretion of propionylglycine in propionic acidaemia, *Clin. Sci.*, 42, 665, 1972.
32. Gompertz, D. and Draffan, G. H., The identification of tiglylglycine in the urine of a child with β-methylcrotonylglycinuria, *Clin. Chim. Acta*, 37, 405, 1972.
33. Greer, M. and Williams, C. C., Diagnosis of branched-chain ketonuria (maple syrup urine disease) by gas chromatography, *Biochem. Med.*, 1, 87, 1967.
34. Hutterer, F., Roboz, J., Sarkozi, L., Ruhig, A., and Bacchin, P., Gas chromatograph—mass spectrometer-computer system for detection and identification of abnormal metabolic products in physiological fluids, *Clin. Chem.*, 17, 789, 1971.
35. Jellum, E., Stokke, O., and Eldjarn, L., Combined use of gas chromatography, mass spectrometry, and computer in diagnosis and studies of metabolic disorders, *Clin. Chem.*, 18, 800, 1972.
36. Kesner, L. and Muntwyler, E., Automatic determination of weak organic acids by partition column chromatography and indicator titration, *Anal. Chem.*, 38, 1164, 1966.
37. Kesner, L., Organic acids in urine. B. Acids of tricarboxylic acid cycle, in *Clinical Biochemistry*, Curtius, H. Ch., Ed., Walter de Gruyter and Co., Berlin, 1973, in press.
38. Ando, T., Rasmussen, K., Wright, J. M., and Nyhan, W. L., Isolation and identification of methylcitrate, a major metabolic product of propionate in patients with propionic acidemia, *J. Biol. Chem.*, 247, 2200, 1972.
39. Humbert, J. R., Hammond, K. B., Hathaway, W. E., Marcoux, J. G., and O'Brien, D., Trimethylaminuria: The fish odour syndrome, *Lancet*, 2, 770, 1970.
40. Morrow, G., III, Schwarz, R. H., Hallock, J. A., and Barness, L. A., Prenatal detection of methylmalonic acidemia, *J. Pediatr.*, 77, 120, 1970.
41. Hansen, S., Perry, T. L., and Lask, D., Urinary bacteria: potential source of some organic acidurias, *Clin. Chim. Acta*, 39, 71, 1972.

Chapter 5

MUCOPOLYSACCHARIDES

INTRODUCTION

Mucopolysaccharidoses consist of a group of genetic disorders in mucopolysaccharide (MPS) metabolism.[1] The clinical and biochemical abnormalities have only recently been delineated, largely with the availability of chemical methods for the determination of mucopolysaccharides and enzyme studies. Methods for the quantitative measurement and characterization of MPS are complicated and laborious. These measurements should be performed at specialized laboratories engaged in research work. Several methods have been described. Urine specimens are first concentrated, dialyzed, or the MPS precipitated. Identification and quantitative measurements of the MPS are then carried out by electrophoresis,[1,2] by determination of uronic acid,[3] or by fractionation on a Sephadex column.[1,4] Description of these quantitative methods is beyond the scope of this book. Several screening tests have been developed for clinical purposes. These tests are generally based upon the metachromatic staining properties or dependent upon the precipitation of MPS by various agents.

There has been considerable confusion about the nomenclature of the MPS. A new system has been suggested by Jeanloz.[5] Both the old and new names for these MPS are listed in Table 5-1 for comparison.

SCREENING TESTS

Toluidine Blue O Spot Test (Berry Spot Test)[6]

Reagents: Toluidine blue O 0.04% in 2% acetic acid, pH2; Ethanol, 95%.

Standard solution: Chondroitin sulfate B 1.0 mg/ml.

Method: A urine specimen of reasonable concentration, such as first morning specimen, is preferable. The specimen should be shaken before spotting to avoid any settling of the MPS.

On a piece of filter paper Whatman No. 1 apply the urine specimen in three separate spots, each containing 5, 10, or 25 μl and one spot 5 μl standard solution. A disposable 5 μl micropipette* is used and the spot is dried between applications. After the spots have been dried by a stream of air, the filter paper is dipped in the toluidine blue solution for about 1 min and rinsed in 95% ethanol.

Results: The urine spot turns purple on a blue background when an increased amount of MPS is present. Traces of purple color in the 25 μl spot have been seen in normal persons. When the 5 μl spot is purple it is definitely abnormal. A normal person excretes less than 20 mg of MPS in 24 hr. Most of these (80 to 90%) are chondroitin sulfates. The remainder consists mainly of dermatan sulfate and heparan sulfate. Infants under 1 year of age excrete more MPS than do older children. Less than 1% of the newborns checked with this test give a false-positive result.[1] In patients with mucopolysaccharidosis, not only are the total amounts of MPS increased, but also the proportions are changed (Table 1-6). In Morquio's syndrome (mucopolysaccharidosis IV) the total amount of MPS may be only twice normal, but the proportion of keratan sulfate is greatly increased. This disorder, therefore, is more likely to give a negative than a positive result with most of the screening tests. If clinical manifestation is highly suggestive of the diagnosis, the qualitative picture of urinary MPS distribution should be studied regardless of the results of the screening test.

Alcian Blue Spot Test[7]

Reagents: Alcian blue, 1 g in 90% glacial acetic acid.

Method: Ten milliliters of urine are applied to a piece of filter paper Whatman No. 1 or No. 4. After the urine is dried, the paper is dipped in the Alcian blue solution. Rinse out the reagent with the water, then wash in glacial acetic acid and again in water.

Results: The urine spot stains bright blue on a white background in the presence of an increased amount of MPS.

---

*Microcaps, Drummond Scientific Co.

TABLE 5-1

Old names	New names
Acid mucopolysaccharides	Glycosaminoglycuronoglycans
Chondroitin sulfate A	Chondroitin-4-sulfate
Chondroitin sulfate B	Dermatan sulfate
Chondroitin sulfate C	Chondroitin-6-sulfate
Heparitin sulfate	Heparan sulfate
Keratosulfate	Keratan sulfate

Acid Albumin Turbidity Test

This test was originally reported by Dorfman and Ott for the assay of hyaluronidase and modified by Carter et al.[8]

Reagents: Phosphate-citrate buffer, 0.3 M, pH 5.5 containing 0.15 M sodium chloride. (Dissolve 80.44 g $Na_2HPO_4 \cdot 7H_2O$, 9.0 g NaCl, and 30.0 g citric acid monohydrate in 900 ml water. Adjust pH and bring volume to 1 liter.

Acid albumin reagent, 0.1% of albumin in 0.1 M acetic buffer, pH 3.75. (Dissolve 13.6 g sodium acetate trihydrate in 800 ml water and add 5.7 ml glacial acetic acid and 1.0 g bovine albumin, fraction V. Adjust pH to 3.75 with 10 N hydrochloric acid. Make up the volume to 1 liter.)

Hydrochloric Acid, 5 N.

Quantitative Turbidimetric Test

Method: All reagents and urine should be at room temperature. To 1.0 ml urine add 1.0 ml distilled water, 0.5 ml of the phosphate-citrate buffer, pH 5.5, and 8.0 ml of the acid-albumin reagent. Mix and wait for 10 min. The turbidity is read at 600 nm against a blank containing the same as above but with distilled water instead of acid albumin reagent.

A standard curve is prepared ranging from 2.0 to 15.0 mg/100 ml aqueous chondroitin sulfate.

Results: A value above 1.0 mg/100 ml chondroitin sulfate equivalent is considered positive. With this cut-off point, Carter et al.[8] found that over 90% of the patients with Hurler's syndrome gave a positive result and that 17.2% of non-Hurler patients gave false-positive results.

Gross Turbidity Test

This is a simplified method of the above turbidimetric test.

Method: To 1 ml of urine, add 2 drops of 5 N hydrochloric acid and 2 ml of acid albumin

reagent. Mix and let stand at room temperature for 10 min. Turbidity is estimated visually and graded 0 to 4+.

Results: Over 90% of the Hurler patients gave 1+ or more, and only 6.3% of the non-Hurler patients gave false-positive results, as reported by Carter et al.[8]

CTAB (Cetyltrimethylammonium bromide) Test[9]

Reagent: Cetyltrimethylammonium bromide 5% (w/v) in 1 M citrate buffer, pH 6.0.

Method: To 5 ml of urine, add 1 ml reagent. Mix well by swirling. Read the turbidity after 30 min.

Both reagent and urine should be at room temperature to avoid a false-positive result.

Result: A heavy, flocculent precipitate is positive.

CPC (Cetylpyridinium chloride) Citrate Turbidity Test

A CPC turbidity test was originally described by Manley and Hawksworth[2] and modified by Pennock[10,11] as follows:

Reagents: a. Citrate buffer, ph 4.8 (Dissolve 9.68 g citric acid and 15.88 g trisodium citrate in 1 l. water). b. Cetylpyridinium chloride 0.1% in the citrate buffer pH 4.8 (w/v). These reagents remain stable for one year at room temperature.

Method: Set up two tubes each containing 1 ml of filtered urine. To one tube add 1 ml CPC reagent (b) and to the other add 1 ml citrate buffer (a). The second tube is used as a blank. After mixing, the tubes are left standing at room temperature for at least 30 min and are then mixed again. The optical density (O.D.) is read at 680 nm. Standard solutions of chondroitin sulfate ranging from 0.5 to 10 mg/100 ml are treated in the same manner.

Results: Results are reported as CPC units/100 ml or CPC units/g creatinine (1 unit = O.D. of 1 mg/100 ml of chondroitin sulfate under the given test condition). Figure 5-1 shows the normal range at different ages.

It is important that specimens and reagents be at room temperature since CPC will precipitate at lower temperatures and will give false-positive results. A reasonably concentrated urine specimen should be used.

Berry Spot Test in Combination with Amino Acid Chromatogram

MPS stays at the origin in the BuAc solvent for paper chromatography of amino acids. Berry et al.[12] described a simple method for performing the toluidine blue O test on a one-dimensional chromatogram for amino acids. After the paper is developed in BuAc and before staining, the section containing the origin where the urine specimen is spotted is cut off and stained with toluidine blue O reagent. The same procedure can be carried out with a chromatogram already treated with ninhydrin.[13] The ninhydrin color fades with the toluidine blue reagent. A purple color at the origin indicates the presence of an increased amount of MSP.

"MPS" Papers* Spot Test

Reagents: a. Squares of paper impregnated with

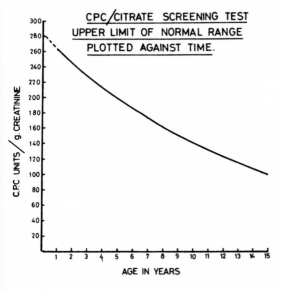

FIGURE 5-1. Urinary glycosaminoglycan excretion in children at various ages. Only the upper limit of normal range is shown. (Courtesy of Dr. C. A. Pennock.)

*Ames Company

Azure A dye are available commercially.* These should be stored in a cool, dry place in the plastic envelope supplied.

b. A wash solution is prepared by mixing 0.1 ml glacial acetic acid, 20 ml anhydrous methanol, and 200 ml distilled water.

Method: With forceps place one test square in a petri dish. Apply 1 drop of urine in the center of the test paper. Wait for 3 min and pour wash solution (b) into the dish, enough to cover the test square. Allow it to stand for 5 min with periodic mild agitation. Using the forceps, place the test square on a piece of clean white filter paper for reading.

Results: The presence of a blue-purple spot on a blue background is a positive result. No spot is visible in normal urine. The sensitivity of this test is 0.02 mg/ml of chondroitin sulfate. Increased amounts of heparin in the urine may cause a false-positive reaction.

COMMENTS

Various tests have been evaluated by different investigators. Some have been found to be effective and dependable by one investigator but not by others. Variations in the test results probably depend to a certain degree upon the exact condition under which the test is performed. Both false-positive and false-negative results have been obtained with each of the tests. Carter et al.[8] found a considerable degree of overlapping of results between normal and Hurler patients by using the toluidine blue spot test. They judged the gross turbidity test to be simple and reliable. The test was positive in 90.7% of the Hurler patients, and the false-positive rate was 6.3% in non-Hurler controls. Pennock et al.[11] observed that both the acid albumin test and the toluidine blue test gave false-negative results. A high proportion of false-positive results but no false-negative results were found with the Alcian blue test. They suggest that the CPC-citrate turbidity test be used for screening, and the CPC-precipitable uronic acid test as confirmatory test. Procopis et al.[14] found the CTAB test to be inferior to the Alcian blue test because it resulted in too many false-positive and false-negative results.

Sabater et al.[15] used a modified toluidine blue spot test to screen large numbers of newborn filter paper urine specimens. There were 0.68% (103 cases) positive results in the 15,000 specimens obtained at about 20 days of life. Repeat urine specimens were received from 78, and 5 of these were still positive at 3 months and 6 months. Three of these five infants had a transient enzyme deficiency which became normal at one year of age. The other two were found to have the enzyme defect, β-galactosidase deficiency, without the clinical manifestation of mucopolysaccharidosis.

The new spot test using a commercial test paper is considered reliable by Berman et al.[16] Rezvani et al.[17] compared this MPS paper test with the gross acid albumin test and the CTAB test and concluded that the gross albumin test is the most suitable screening because it gives only 2% false-positive and no false-negative results. The MPS paper test gives 31% false-positive results. However, if the weakly positive tests are dismissed,

this test could be considered comparable to the gross acid albumin test.

In summary, the reports in the literature regarding the efficiency of screening tests for mucopolysaccharidoses are inconclusive. No test is considered to be the best. Thus, it appears advantageous to set up two simple screening tests: one is the spot test based upon the metachromatic staining property of the MPS, such as "Berry Spot" Test (toluidine blue test)[6] or MPS paper test,[16] and the other is a turbidity or precipitation test. The gross acid albumin test[8] is probably the simplest and most reliable. A positive result with one test can be confirmed by the other test. Positive results with both tests would strongly suggest the presence of increased MPS in the urine. A definitive diagnosis can be obtained by quantitative measurement of the MPS and by establishing the absence of activity of the involved enzyme in the blood, urine, or cultured cells.

REFERENCES

1. Dorfman, A. and Matalon, R., The mucopolysaccharidoses, in *The Metabolic Basis of Inherited Disease*, 3rd ed., Stanbury, J. B., Wyngaarden, J. B., and Fredrickson, D. S., McGraw-Hill, New York, 1972, 1218.
2. Manley, G. and Hawksworth, J., Diagnosis of Hurler's Syndrome in the hospital laboratory and the determination of its genetic type, *Arch. Dis. Child.*, 41, 91, 1966.
3. Teller, W. M., Rosevear, J. W., and Burke, E. C., Identification of heterozygous carriers of gargoylism, *Proc. Soc. Exp. Biol. Med.*, 108, 276, 1961.
4. DiFerrante, N., The measurement of urinary mucopolysaccharides, *Anal. Biochem.*, 21, 98, 1967.
5. Jeanloz, R. W., The nomenclature of mucopolysaccharides, *Arthritis Rheum.*, 3, 233, 1960.
6. Berry, H. K. and Spinanger, J., A paper spot test useful in study of Hurler's syndrome, *J. Lab. Clin. Med.*, 55, 136, 1960.
7. Carson, N. A. J. and Neill, D. W., Metabolic abnormalities detected in a survey of mentally backward individuals of Northern Ireland, *Arch. Dis. Child.*, 37, 505, 1962.
8. Carter, C. H., Wan, A. T., and Carpenter, D. G., Commonly used tests in the detection of Hurler's syndrome, *J. Pediatr.*, 73, 217, 1968.
9. Renuart, A. W., Screening of inborn errors of metabolism associated with mental deficiency or neurological disorders or both, *N. Engl. J. Med.*, 274, 384, 1966.
10. Pennock, C. A., A modified screening test for glycosaminoglycan excretion, *J. Clin. Pathol.*, 22, 379, 1969.
11. Pennock, C. A., Mott, M. G., and Batstone, G. F., Screening for mucopolysaccharidoses, *Clin. Chim. Acta*, 27, 93, 1970.
12. Berry, H. K., Leonard, C., Peters, H., Granger, M., and Chunekamrai, N., Detection of metabolic disorders. Chromatographic procedures and interpretation of results, *Clin. Chem.*, 14, 1033, 1968.
13. Curtis, H. T. and Buist, N. R., Use of toluidine blue for sequential staining of urinary amino acid chromatograms, *J. Chromatogr.*, 57, 165, 1971.
14. Procopis, P. G., Turner, B., Ruxton, J. T., and Brown, D. A., Screening tests for mucopolysaccharidosis, *J. Ment. Defic. Res.*, 12, 13, 1968.
15. Sabater, J., Villalba, M., and Maya, A., Mass screening for newborns for mucopolysaccharidoses, *Abstracts, 8th Int. Cong. Clin. Chem.*, Copenhagen, 1972.
16. Berman, E. R., Vered, J., and Bach, G., A reliable spot test for mucopolysaccharidoses, *Clin. Chem.*, 17, 886, 1971.
17. Rezvani, I., Collipp, P. J., DiGeorge, A. M., and Punnett, H. H., Screening tests for detection of mucopolysaccharide disorders: evaluation of a new test, *Pediatr. Res.*, 6, 401, 1972.

Chapter 6

MISCELLANEOUS DISORDERS

Table 6-1 lists a group of miscellaneous metabolic disorders, with the laboratory techniques for their detection referenced. Details of the techniques are not described. The tests listed here are rarely indicated in a screening laboratory, but some of them are performed when a specific screening program is in operation. For instance, pilot screening programs for the detection of Tay-Sachs disease (heterozygotes), sickle cell disease (homozygotes and heterozygotes), glucose-6-phosphate dehydrogenase deficiency, and alpha$_1$-antitrypsin deficiency are in operation in various cities, but the value of such programs, except that for Tay-Sachs disease, has been widely debated.[16]

It is noteworthy that orotic acid is excreted in large quantity by patients with ornithine carbamyltransferase (OCT) deficiency. While in-creased glutamine is the only amino acid abnormality in this disorder, the finding of orotic acid and other pyrimidine derivatives, such as uracil and uridine, are important clues leading to the investigation of the urea cycle enzymes.[11] For the diagnosis of metachromatic leukodystrophy, specific arylsulfatase A measurements have largely replaced the earlier screening tests described by Austin demonstrating metachromatic bodies and excess sulfatide ("fluff test") in the urine sediment.[4] Certain precautions should be taken when setting up this and the other listed enzyme assays. Each laboratory must establish its own optimum assay conditions and control values.

When diagnostic tests are only occasionally required, they should be performed at laboratories with a special interest in these disorders to insure reliability of the results.

TABLE 6-1

Laboratory Detection of Miscellaneous Metabolic Disorders

Disorders	Test	References
Tay-Sachs disease	Hexosaminidase A	1-3
Sulfatide lipidosis (Metachromatic leukodystrophy)	Arylsulfatase A sulfatide	4
Phytanic acid storage disease (Refsum's syndrome)	Phytanic acid	5, 6
Lesch-Nyhan syndrome (Hypoxanthine-guanine phosphori-bosyl transferase deficiency)	Uric acid Hypoxanthine-guanine phosphoribosyl trans-ferase	7, 8
Orotic aciduria Ornithine carbamyl-transferase deficiency	Orotic acid	9-11
Sickle cell disease	Hemoglobin S	12
Glucose-6-phosphate dehydrogenase deficiency	Glucose-6-phosphate dehydrogenase	13, 14
Alpha$_1$-antitrypsin deficiency	Alpha$_1$-antitrypsin	Dr. William H. Murphey, Children's Hospital, Buffalo, N.Y.
Primary hyperoxalurias	Oxalic acid	15

REFERENCES

1. O'Brien, J. S., Okada, S., Chen, A., and Fillerup, D. L., Tay-Sachs disease. Detection of heterozygotes and homozygotes by serum hexosaminidase assay, *N. Engl. J. Med.*, 283, 15, 1970.
2. Kaback, M. M. and Zeiger, R. S., Heterozygote detection for Tay-Sachs disease (TSD) in a sample American-Jewish population, *Pediatr. Res.*, 6, 362, 1972.
3. Cotlier, E., Tay-Sachs disease: abbreviated serum hexosaminidase. A test from finger-tip samples, *Clin. Chim. Acta*, 38, 233, 1972.
4. Moser, H. W., Sulfatide lipidosis: Metachromatic leukodystrophy, in *The Metabolic Basis of Inherited Disease*, 3rd ed., Stanbury, J. B., Wyngaarden, J. B., and Fredrickson, D. S., Eds., McGraw-Hill, New York, 1972, 688.
5. Try, K., Heredopathia atactica polyneuritiformis (Refsum's disease), the diagnostic value of phytanic acid determination in serum lipids, *Eur. Neurol.*, 2, 1, 1969.
6. Steinberg, D., Avigan, J., Mize, C. E., Baxter, J. H., Cammermeyer, J., Fales, H. M., and Highet, P. F., Effects of dietary phytol and phytanic acid in animals, *J. Lipid Res.*, 7, 684, 1966.
7. McInnes, R., Lamm, P., Clow, C. L., and Scriver, C. R., A filter paper sampling method for the uric acid:creatinine ratio in urine. Normal values in the newborn, *Pediatrics*, 49, 80, 1972.
8. Seegmiller, J. E., Rosenbloom, F. M., and Kelley, W. N., Enzyme defect associated with a sex-linked human neurological disorder and excessive purine synthesis, *Science*, 155, 1682, 1967.
9. Rogers, L. E. and Porter, F. S., Hereditary orotic aciduria. II. A urinary screening test, *Pediatrics*, 42, 423, 1968.
10. Efron, M. L., High voltage paper electrophoresis, in *Chromatographic and Electrophoretic Techniques, Volume II, Zone Electrophoresis*, 3rd ed., Smith, I., Ed., John Wiley & Sons, New York, 1968, 166.
11. Levin, B., Oberholzer, V. G., and Sinclair, L., Biochemical investigations of hyperammonaemia, *Lancet*, 2, 170, 1969.
12. Hosty, T. S., Tomlin, G. A., and Atkins, R. J., Testing for hemoglobinopathies in public health laboratories, *Health Lab. Sci.*, 1972.
13. Beutler, E., A series of new screening procedures for pyruvate kinase deficiency, glucose-6-phosphate dehydrogenase deficiency, and glutathione reductase deficiency, *Blood*, 28, 553, 1966.
14. Murphey, W. H., Patchen, L., and Guthrie, R., Screening tests for argininosuccinic aciduria, orotic aciduria, and other inherited enzyme deficiencies using dried blood specimens, *Biochem. Genet.*, 6, 51, 1972.
15. Hodgkinson, A., Determination of oxalic acid in biological material, *Clin. Chem.*, 16, 547, 1970.
16. Levy, H. L., Genetic screening, in *Advances in Human Genetics*, Vol. 4, Harris, H. and Hirschhorn, K., Eds., Plenum Press, New York, in press.

Chapter 7

STRUCTURAL GUIDELINES FOR
METABOLIC DISORDER SCREENING PROGRAM

Two types of laboratories are currently involved in the detection of genetic metabolic disorders: the hospital or patient-oriented laboratory and the public health or newborn-oriented laboratory. The patient-oriented laboratory is usually the outgrowth of a research laboratory which is engaged in the study of inborn errors of metabolism and in the treatment and clinical management of these patients. It is this type of laboratory which initially conducted surveys in institutions for mentally retarded patients. Its main function is to serve as a diagnostic and patient-care facility and, in addition, to discover "new" metabolic disorders.

The newborn-oriented screening laboratory is set up in the interest of the general population. Its aim is early diagnosis and treatment in order to prevent clinical disabilities, primarily mental retardation. Of equal importance is an understanding of the natural history and incidence of these disorders so that the proper therapy can be instituted and appropriate genetic counseling given. Most of the laboratories engaged in mass newborn screening are state public health laboratories. Phenylketonuria (PKU) was the first disorder tested on a large scale in a newborn screening program. With the obvious success of the PKU screening programs in many states[*] and countries, the newborn screening program has been expanded to include screening for many other metabolic disorders at very little additional cost. By far, the most comprehensive program is the one conducted in Massachusetts.

Aberrations in amino acid and sugar metabolism have been studied extensively, while screening for disorders in mucopolysaccharide and lipid metabolism has been performed only on a small scale, largely because of the lack of simple and effective laboratory techniques and partly because there is no effective treatment for these conditions and early diagnosis is not imperative.

HOSPITAL OR PATIENT-ORIENTED METABOLIC SCREENING LABORATORY

Specimens are usually received from clinicians who wish to rule out hereditary metabolic disorders as the cause of the patient's clinical problems. Since the manifestations of metabolic disorders may vary widely and those of the undiscovered disorders are not known, patients with any unsolved problems deserve a screening test. At some hospitals, metabolic screening is included in the admission routine work-up of all pediatric patients.

It is important to provide some clinical information on the requisition slip since many diseases have characteristic symptoms and signs, and since factors such as age can affect the excretion of amino acids and mucopolysaccharides. Moreover, the laboratory can perform other pertinent tests in addition to the routine procedures that will make screening more meaningful.

Table 7-1 lists all the tests suggested for routine metabolic screening. Random urine and blood specimens on filter paper are adequate for general screening, as has been previously discussed. When the clinical diagnosis strongly suggests an organic acid abnormality such as propionic acidemia or isovaleric acidemia, an additional GLC analysis of the serum is indicated. When an unusual finding is obtained by routine screening, it should be followed up by studying a repeat specimen at least three days after the patient has stopped all medication to rule out any exogenous cause. A 24-hr or a timed urine collection and a serum or plasma sample are required for further study.

When the diagnosis of a metabolic disorder is made in one member of the family, a survey of siblings and parents is indicated. Other relatives should be included when their medical history suggests that they may be similarly affected. It should be emphasized that when all siblings are

---

*As of November, 1970, 43 states have laws for mandatory PKU testing.[1]

TABLE 7-1

Suggested Procedures for Metabolic Screening

Disorders	Urine			Blood	
	Routine tests				
	Chemical or spot test	Chromatography	Further study	Routine test	Further study
Amino acids	Ferric chloride Dinitrophenylhydrazine Cyanide-nitroprusside Reducing substances Thiocyanate	Paper, in combination with high voltage electrophoresis	Column chromatography; gas-liquid chromatography	Paper chromatography	Column chromatography; gas-liquid chromatography; ammonia determination
Sugars	Reducing substances Anthrone	None	Paper chromatography	None	Enzyme assays (e.g., Beutler fluorescence test)
Organic acids	Odor Methylmalonic acid Dinitrophenylhydrazine	Paper	Quantitative measurement by chemical methods, gas-liquid chromatography, and partition chromatography	Gas-liquid chromatography	Gas-liquid chromatography
Mucopoly- saccharides	Berry spot Albumin turbidity	None	Column chromatography; uronic acid determination; enzyme assays	None	Enzyme assays

mentally retarded and no biochemical abnormality can be found in any of them, it is important to screen the parents, particularly the mother, for any metabolic abnormalities since maternal PKU as a cause of mental retardation[2] and congenital anomalies[3] in offspring is well recognized.

PUBLIC HEALTH-ORIENTED OR NEWBORN SCREENING PROGRAM

Based upon recommendations of the Committee on Fetus and Newborn of the American Academy of Pediatrics for the screening of newborn infants for metabolic disease,[4] the Maternal and Child Health Service of the United States Department of Health, Education, and Welfare suggests the following criteria in developing a PKU screening program:[5]

1. A test should be chosen which is of proven efficiency as a screening tool. It should permit simplicity and economy in the collection, storage, and determination of samples.

2. The testing of the samples should be done by a laboratory with adequate facilities which handles a sufficient volume of samples to maintain skill in recognizing abnormal findings. A system of quality control developed on a statewide or regional basis would insure reliability of results.

3. The initial test should be applicable before newborn infants are discharged from the nursery and thereby become less accessible.

4. A follow-up screening test on blood or urine may be desirable at a few weeks of age.

5. All presumptive positive cases should be checked by a specific confirmatory test for an elevated blood phenylalanine level.

6. Adequate procedures for referral of infants to evaluation and treatment centers should be an integral part of the program.

These criteria can also be applied to the development of an expanded screening program, with one addition. When determining which disorders to include in the screening, those in which the patients will benefit most from early recognition should be considered. In those diseases where the clinical symptoms or cause cannot be alleviated or altered by treatment, a program for screening for detection of the carriers and of genetic counseling should be developed.

Type of Disorders

Screening for a disease is possible when the test is simple and economical. With limited resources and manpower, diseases that cause serious clinical abnormalities and for which treatment is available should be given high priority. PKU, the most common amino acid metabolic disorder, and other similar amino acid metabolic disorders, as well as the disorders in sugar metabolism and organic acid metabolism, all meet these criteria and have been screened at some laboratories. It can be predicted that at many laboratories, an expanded program including the screening for all these disorders will be forthcoming.

Advances in enzyme assays have made possible the detection of carriers for certain hereditary disorders such as Tay-Sachs disease. Tay-Sachs disease is a fatal degenerative disease of the nervous system found mainly in Jewish children. There is no effective treatment for this condition. However, since a prenatal diagnosis can be made, a program for the detection of the carrier and genetic counseling is being evaluated to determine its success as a preventive measure. A pilot screening program to test Jewish couples of child-bearing age on a voluntary basis is being carried out in the Baltimore–Washington, D.C. area,[6] and proper genetic counseling is being offered to those at risk.

There are also pilot screening programs for sickle cell disease which is most common among black people and causes severe clinical disability and early death. Symptomatic treatment is possible. Carriers for sickle trait can be detected by hemoglobin electrophoresis. Various investigators are in the process of developing and testing techniques of hemoglobin electrophoresis applicable to mass screening and sensitive enough to detect small amounts of sickle hemoglobin in young infants to determine their efficacy in the diagnosis of sickle cell disease. Proper genetic counseling to carriers, early medical attention to patients, and possible prenatal diagnosis will benefit those individuals at risk.

Techniques

Techniques suitable for mass screening should be simple, automated if possible, and economical. The advantages and disadvantages of several techniques that have been field-tested or considered in mass screening are discussed here.

Bacterial Inhibition Assays

At the present time, most of the screening laboratories use Guthrie bacterial inhibition assay (GBIA) for the detection of hyperphenylalaninemia.[7] Other GBIA for methionine (for the detection of hypermethioninemia in cystathionine synthase deficiency), for leucine (for the detection of maple syrup urine disease), for histidine (for the detection of histidinemia), and for tyrosine (for the detection of hereditary tyrosinemia) have been used in several laboratories. Detection of methionine and leucine by the GBIA can be justified since the amino acid abnormalities in cystathionine synthase deficiency and in maple syrup urine disease are prominent in the blood; early treatment for these disorders is advisable. However, the detection of histidine and tyrosine by GBIA has been the subject of some controversy. Although histidinemia has been effectively diagnosed with the GBIA method,[9] it can also be easily detected by urine screening at four to six weeks of age using paper chromatography.[10] Recent experiences with this disorder have indicated that it is a benign condition and not associated with any specific neurologic abnormality.[10] Setting up a specific test for an abnormality which causes no clinical disability does not seem to be justified.

The value of testing for tyrosine elevation is also questionable since more than one million newborn infants have been tested[7] and not one case of hereditary tyrosinosis or tyrosinemia has been discovered. These findings can be interpreted in the following ways: (1) the diagnosis of hereditary tyrosinemia has been missed; (2) tyrosine elevation in hereditary tyrosinemia is either not prominent or is absent at the age of testing (usually first week of life); (3) hereditary tyrosinemia is such a rare disease in the general population that testing for it is not warranted. In any case, the usefulness of testing for blood tyrosine in newborns as a means of detecting hereditary tyrosinemia should be reappraised.

The most frequent finding in newborn screening is tyrosine elevation due to the transient type of neonatal tyrosinemia. This type of tyrosinemia has been considered to be a benign entity, and no clinical abnormality has been noted at the time when tyrosinemia is present[11] or at the time of short-term follow-up.[12,13] However, a recent report of a long-term study by Menkes et al.[14] has shown that a high tyrosine level in the neonatal period may have an adverse effect on the intellect of large low-birthweight (2 to 2.5 kg) infants. They observed a significant difference in the intellectual performance at the age of seven to eight years between the group who had high blood tyrosine levels (15 mg/100 ml or higher) and the group who had low tyrosine levels (<15 mg/ml) in the neonatal period. Interestingly, the difference was not observed in premature infants weighing less than 2 kg at birth. If the causal relationship between high blood tyrosine level and impairment of intellectual performance can be proven by additional long-term follow-up studies, screening newborns for tyrosine elevation would be warranted.

It should be mentioned that in a PKU screening program, tyrosine measurement is very important because hyperphenylalaninemia secondary to hypertyrosinemia should be ruled out before the diagnosis of a defect in phenylalanine hydroxylation (PKU) per se can be made. For this purpose, tyrosine determination by a fluorometric method or chromatographic methods is more convenient than GBIA.

Of the two microbiological assays for galactose, the Paigen bacteriophage test has certain advantages over the Guthrie metabolite inhibition assay (cf., Chapter 3); the former is recommended for galactose screening.

Chromatographic Method

The two most widely used paper chromatographic methods for amino acids are those of Efron et al.[15] and of Scriver et al.[16] (cf., Chapter 2); both have been extensively applied to screening and proven satisfactory. These techniques are not as sensitive as the GBIA for detecting a mild elevation (<5 mg/100 ml). Because the GBIA are designed to detect only a few of the amino acids, a pilot study for additional testing of blood by paper chromatography was conducted in Massachusetts to determine whether chromatography was better; no additional cases were found. Raine et al.[17] recently reported their 3-year experience of screening more than 55,000

newborns using plasma chromatography as described by Scriver. Among the disorders diagnosed, hyperprolinemia is the only one that cannot be detected by GBIA. Therefore, it is fair to say that GBIA for phenylalanine, methionine, and leucine are adequate for newborn blood screening in cases where urine screening will follow. Paper chromatography can be used as a confirmatory test for the elevation detected by GBIA and as a test for any other associated abnormality. For the same reason, TLC of the blood has no real advantage over GBIA in newborn blood screening, but it too is a valuable confirmatory test.[18] It is faster but more expensive than paper chromatography.

Paper chromatography is generally the most valuable technique for examining the whole amino acid or organic acid pattern in urine. One-dimensional separation as the preliminary routine test followed by two-dimensional separation in cases with suspected abnormalities has been used effectively in the detection of various amino acid metabolic and transport abnormalities.[19]

Ion-exchange chromatography for amino acids and GLC for organic acids are both suitable as back-up techniques to confirm a presumptive finding noted with GBIA or paper chromatography and to measure quantitatively the abnormal compound in a follow-up specimen.

Enzyme Assays

Enzyme assays using filter paper blood specimens have one obvious advantage, that is, the test can be done on cord blood. The two diseases for which enzyme assays are available, argininosuccinic aciduria (cf., Chapter 2) and classical galactosemia (Gal-1-P uridyltransferase deficiency) (cf., Chapter 3) may both result in a fulminating fatal neonatal course. Affected infants may die undiagnosed, and these occurrences are probably more common than we know. Therefore, the inclusion of these assays in routine cord blood testing is advisable. The Beutler fluorescence test for transferase activity is also valuable as a differential diagnostic test when galactose elevation is found.

Fluorometric and Spectrophotometric Techniques

Fluorometric methods for quantitative measurement of phenylalanine, tyrosine, and histidine have been developed and modified for automated use on filter paper blood specimens (cf., Chapter 2). The fluorometric method can be applied when screening is limited to phenylalanine, but it does not seem practical when routine testing includes other amino acids such as leucine and methionine since all of these can be done conveniently by GBIA. These fluorometric tests are more useful as a second method for quantitative measurement of phenylalanine and tyrosine than as a routine screening test; however, this technique does not have as much versatility as the amino acid analyzer which can be used to measure all other amino acids.

The spectrophotometric method for phenylalanine, tyrosine, and tryptophan described by LaDu and Michael[20] is technically difficult. Extreme care must be taken to avoid erroneous results.

Types of Specimens

It is recommended that both blood and urine be screened.

Blood specimens are used for the detection of enzyme defects as well as the accumulation of abnormal metabolites. Classical galactosemia should be diagnosed and treated as early as possible. The Beutler fluorescence test detects the enzyme defect and the Paigen bacteriophage test for galactose is sensitive enough to detect mild elevations in the first few days of life. In certain amino acid disorders, there are greatly increased amino acid concentrations in the blood (PKU, maple syrup urine disease, and cystathionine synthase deficiency), but only mild changes in the urine in the first month of life; therefore, blood specimens should be used for screening for these disorders. On the other hand, urine testing is much better for those disorders in which there are marked changes in the urine and for the group of transport disorders in which no abnormality is found in the blood.

Ease in collecting and transporting specimens is an important factor in the success of any screening program. All tests for screening have been so modified that dried specimens of blood or urine on filter paper can be used. These filter paper specimens can be collected with little effort by a nurse or by the parents and can be sent to a central laboratory by ordinary mail. The capillary serum samples used in the paper chromatographic method developed by Scriver et al.[16] can also be collected from a heel puncture and mailed.

Age of Testing

The appropriate age for screening is determined by several factors.

From the practical point of view, blood sampling should be done before discharge from the nursery. In countries such as Great Britain it is common for babies to be born at home, in which case blood can be collected by health visitors on the first visit which would normally be during the second or third week of life.[21] In most instances, the screening test is based upon finding an abnormal amount of a metabolite; therefore, the blood should be taken after the infant has had at least 12 milk feedings. In the U.S. babies are usually discharged from the nursery between three and five days of age, at which time the blood specimen can be obtained. If the infant is discharged prior to this time, a second blood specimen should be obtained preferably at one or two weeks of life, possibly by a health visitor,[21] or at the first well-baby checkup.

For a premature infant remaining in the nursery for an extended period of time, the blood sampling can be done at about two weeks of age and again at discharge. When a disorder can be diagnosed by demonstrating the enzyme deficiency, cord blood specimen is preferable. Cord blood is also useful in studying maternal disease. For instance, the cord blood phenylalanine concentration is greatly elevated in an infant born of a PKU mother, and it drops rapidly to a normal range within 48 hr.[22]* Therefore, the diagnosis of maternal PKU can only be made by testing cord blood and not from blood obtained at four or five days or age.**

Urine specimens are collected when the infant is three to six weeks of age. This age is chosen somewhat arbitrarily and is subject to change if new information becomes available. Mothers are given a urine collection kit which includes an information card to which the filter paper for the urine specimen is attached, an instruction sheet for the parents (see appendix), and a return envelope. On the information card, in addition to individual identification, the name and telephone number of

---

*Diagnosis by a ferric chloride (or Phenistix®) urine test during a prenatal visit is preferable since it will permit dietary therapy if necessary. With an increasing awareness of the deleterious effects of maternal PKU on the fetus, this test has been advocated as a routine prenatal procedure.[23]

**Occasionally mild elevation may be observed.

the physician, the type of formula or feeding, and the ointment used are also requested. The last two items provide pertinent information for the interpretation of chromatograms (see discussion on artifact in Chapter 2).

High-risk Baby

Siblings of a child affected with a hereditary metabolic disorder are at high risk with a one-in-four chance of being similarly affected. When such a high-risk infant is born, the screening laboratory should be notified. When the activity of the involved enzyme can be measured, an assay should be performed on cord blood for definitive diagnosis. Otherwise, quantitative measurement for the specific metabolite should be performed on cord blood, and the blood and urine specimens obtained at three days, one week, and two weeks of age. It is uncertain whether those diseases which have not yet been detected by newborn screening can actually be identified in the first several weeks of life. Under these circumstances, the follow-up period should be extended to one or two months, or beyond the age when clinical symptoms usually appear.

Family Survey

Since the disorders are hereditary in origin, the finding of an affected individual should lead to a survey of all members of the family. All of the disorders currently being screened are transmitted as autosomal recessive characters, and there is a one-in-four chance that a sibling will be similarly affected. In the case of a relatively common defect, there is a possibility that the parent may also be affected. Cystinuria is an incomplete recessive trait; the carrier of this trait can be detected by the screening method. When a heterozygous infant is discovered, there is the possibility of homozygous siblings if both parents carry the mutant gene or of heterozygous siblings if only one parent carries the mutant gene. A third possibility is that one parent may be a homozygote. In the Massachusetts screening program we have indeed found cystinuria and iminoglycinuria in both the parents and the children. Hyperprolinemia[24,25] and his-tidinemia[26] have also been reported in two successive generations. Suffice it to say that screening is not complete without family study.

Procedures

When specimens are received at the laboratory, they are put through a battery of routine tests. The procedures of specimen handling used in the Massachusetts screening program, as illustrated in Figures 7-1 and 7-2, are given as reference.

Record Keeping

It can be anticipated that a large volume of data will be collected in a relatively short period in a program of mass screening. In order to make this information useful, it should be carefully cross-indexed and computerized for easy retrieval.

Yield of Newborn Screening and Cost

Valuable information about the genetics, clinical manifestations, and incidence of hereditary metabolic disorders has been obtained by mass screening in the unselected newborn population. A comprehensive review of the screening results has recently been prepared by Levy.[7] Table 7-2 summarizes the incidence of various disorders compiled from data obtained from screening programs in different countries.

The cost of testing one infant varies with the size of the program, and the types and number of tests performed, the fewer tests done at a laboratory, the higher the cost. Again, we will use the Massachusetts program as an example of a laboratory that tests 80,000 to 100,000 infants per year.[7] Almost all infants born in Massachusetts (99% or better) have a screening test on blood specimens obtained at discharge from the nursery (mandatory by law); approximately 80% have urine screening and approximately 25% have a cord blood screening test. Total tests performed include five Guthrie bacterial inhibition assays, the Beutler fluorescence test, the Paigen bacteriophage test, and urine paper chromatographic studies. The cost is approximately $2.00 per infant tested (for both blood and urine). Comparison of the cost benefits of different programs has been made by Levy[7] and by Cunningham.[27]

Patient Referral Facility

A facility to which patients can be referred for further study, treatment, and genetic counseling is an integral part of a screening program, without which the purpose of screening would be defeated. A detailed discussion of the medical follow-up of these patients is beyond the scope of this book.

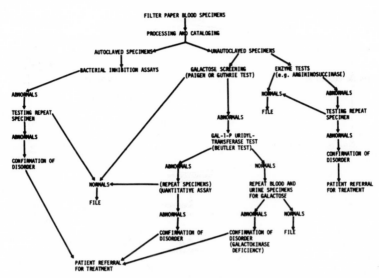

FIGURE 7-1. Diagram of procedures used in screening blood specimens for genetic disorders in Massachusetts. (Shih and Levy, unpublished.)

TABLE 7-2

Incidence of Amino Acid Metabolic Disorders (Data from Various Newborn Screening Programs)[a]

Disorders	Total number tested	Total number found	Incidence
Phenylketonuria	13,538,912[b]	1,186	1:11,500
Maple syrup urine disease	3,245,631[b]	15	1:220,000
Hypermethioninemia (and homocystinuria)	2,654,310[b]	12	1:220,000
Hereditary tyrosinemia	1,401,777[b]	0	0
Histidinemia	1,001,193[c]	41	1:24,000
Hyperprolinemia	200,000[c]	7[e]	
Hyperlysinemia	245,379[d]	1	1:245,000
Hyperglycinemias			
Ketotic	245,379[d]	1	1:245,000
Nonketotic	245,379[d]	1	1:245,000
Argininosuccinic aciduria	245,379[d]	1	1:245,000
Hyperornithinemia	245,379[d]	1	1:245,000
Hartnup disease	554,301[d]	21	1:26,000
Cystinuria, homozygote	554,301[d]	77	1:7,000
Iminoglycinuria	554,301[d]	28	1:20,000
Fanconi syndrome	245,379[d]	1	1:245,000

[a]Table derived from data compiled by Levy.[7]
[b]Blood screening only.
[c]A total of blood and urine screening.
[d]Urine screening only.
[e]Only three have been studied in detail.

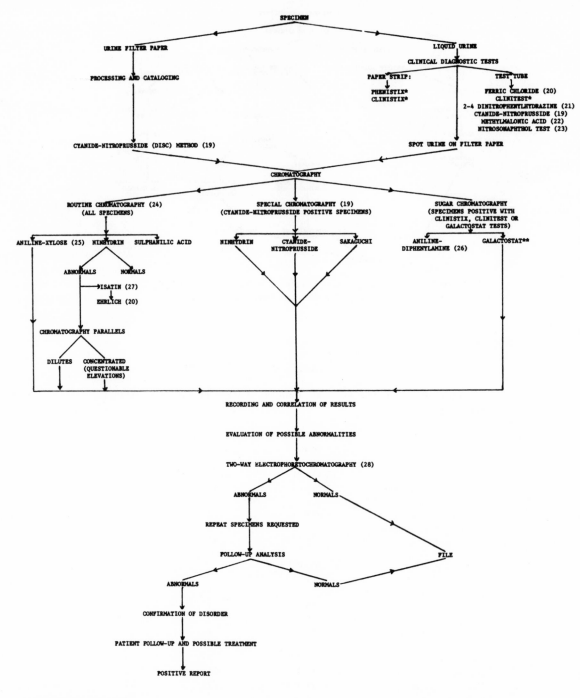

FIGURE 7-2. Diagram of procedures used in screening urines for genetic disorders in Massachusetts. (Reproduced from Levy, H. L. et al., Massachusetts metabolic disorders screening program. I. Technics and results of urine screening, *Pediatrics,* 49, 825, 1972, with permission from the authors and publisher.)

APPENDIX

INSTRUCTIONS FOR PARENTS REGARDING COLLECTION OF FILTER PAPER URINE SPECIMENS

1. Enclosed you will find a printed white filter paper form. Using a *pencil*, please fill out this form completely. (*Please do not use ink*, as it interferes with the test results.)

2. Please tear off (along the dotted line) the right hand portion of the form. This portion is then divided into two pieces.

3. When your baby is about three to four weeks old, simply soak one of the two pieces of the filter paper with urine in either one of two ways:

A. Remove a wet diaper from your baby. Take a wet but unsoiled portion of the diaper and sandwich the filter paper between two folds. Then blot the wet diaper into the filter paper until the paper is thoroughly wet. Then allow the paper to dry, or

B. Place the filter paper inside the baby's diaper. Be sure to put the filter paper in the folds of the diaper where it is not likely to be stained by the baby's stools, because it is urine that we need.

4. The best time to obtain this specimen is before the baby is fed.

5. It is important that the filter paper not be soiled by either the baby's *stool* or by such contaminants as diaper *cremes or oils*. Therefore, please see that no stool gets on the paper and the baby has no diaper creme or oil on at the time the specimen is taken.

6. Attach the impregnated filter paper to the information form and return in the addressed envelope. The additional filter papers are for your convenience in case the first filter paper is contaminated with stool or is otherwise ruined.

7. If the test is normal, you will not hear from us further. If necessary, we shall contact your physician directly.

8. If your baby is being cared for by a well-baby clinic, we shall contact the clinic, if necessary.

9. This service is tax-supported. Thank you for your cooperation.

REFERENCES

1. State Laws Pertaining to Phenylketonuria as of November, 1970, United States Department of Health, Education, and Welfare, Public Health Service, Health Services and Mental Health Administration (Maternal and Child Health Service), U.S. Govt. Print. Off., Washington, D.C., 1971.

2. Mabry, C. C., Denniston, J. C., Nelson, T. L., and Son, C. D., Maternal phenylketonuria. A cause of mental retardation in children without the metabolic defect, *N. Engl. J. Med.*, 269, 1404, 1963.

3. Stevenson, R. E. and Huntley, C. C., Congenital malformations in offspring of phenylketonuric mothers, *Pediatrics*, 40, 33, 1967.

4. Committee on Fetus and Newborn, Screening of newborn infants for metabolic disease, *Pediatrics*, 35, 499, 1965.

5. Recommended Guidelines for PKU Programs for the Newborn, U.S. Department of Health, Education, and Welfare, Maternal and Child Health Service, U.S. Govt. Print. Off., Washington, D.C., 1971.

6. Kaback, M. M., Zeiger, R. S., and Gershowitz, H., A pilot program in the control of genetic disease, *Pediatr. Res.*, 6, 358, 1972.

7. Levy, H. L., Genetic screening, in *Advances in Human Genetics*, Vol. 4, Harris, H. and Hirschhorn, K., Eds., Plenum Press, New York, in press.

8. Kazazian, H. H., Jr., Kaback, M. M., and Nersesian, W. S., Sickle cell hemoglobin production in an aborted midtrimester fetus, *Pediatr. Res.*, 6, 358, 1972.

9. Thalhammer, O., Scheibenreiter, S., and Pantlitschko, M., Histidinemia: detection by routine newborn screening and biochemical observations on three unrelated cases, *Z. Kinderheilkd.*, 109, 279, 1971.

10. Levy, H. L., Shih, V. E., and MacCready, R. A., Inborn errors of metabolism and transport: prenatal and neonatal diagnosis, in *Proceedings of The International Congress of Pediatrics, Volume 4, Genetics*, Vienna, 1971, 1.

11. Avery, M. E., Clow, C. L., Menkes, J. H., Ramos, A., Scriver, C. R., Stern, L., and Wasserman, B. P., Transient tyrosinemia of the newborn: Dietary and clinical aspects, *Pediatrics*, 39, 378, 1967.

12. Menkes, J. H., Chernick, V., and Ringel, B., Effect of elevated blood tyrosine on subsequent intellectual development of premature infants, *J. Pediatr.*, 69, 583, 1966.

13. Partington, M. W., Delahaye, D. J., Masotti, R. E., Read, J. H., and Roberts, B., Neonatal tyrosinemia: A follow-up study, *Arch. Dis. Child.*, 43, 195, 1968.

14. Menkes, J. H., Welcher, D. W., Levi, H., Dallas, J., and Gretsky, N. E., Relationship of elevated blood tyrosine to the ultimate intellectual performance of premature infants, *Pediatrics,* 49, 218, 1972.
15. Efron, M. L., Young, D., Moser, H. W., and MacCready, R. A., A simple chromatographic screening test for the detection of disorders of amino acid metabolism. A technic using whole blood or urine collected on filter paper, *N. Engl. J. Med.,* 270, 1378, 1964.
16. Scriver, C. R., Davies, E., and Cullen, A. M., Application of a simple method to the screening of plasma for a variety of aminoacidopathies, *Lancet,* 2, 230, 1964.
17. Raine, D. N., Cooke, J. R., Andrews, W. A., and Mahon, D. F., Screening for inherited metabolic disease by plasma chromatography (Scriver) in a large city, *Br. Med. J.,* 2, 7, 1972.
18. Schön, R. and Thalhammer, O., 25000 routinemässige Dünnschichtchromatographien bei Neugeborenen. Ergebnisse und Vergleich mit Guthrie-Testen, *Z. Kinderheilkd.,* 111, 223, 1971.
19. Levy, H. L., Madigan, P. M., and Shih, V. E., Massachusetts metabolic disorders screening program. I. Technics and results of urine screening, *Pediatrics,* 49, 825, 1972.
20. LaDu, B. N. and Michael, P., An enzymatic spectrophotometric method for the determination of phenylalanine in blood, *J. Lab. Clin. Med.,* 55, 491, 1960.
21. Komrower, G. M., Fowler, B., Griffiths, M. J., and Lambert, A., A prospective community survey for aminoacidaemias, *Proc. Roy. Soc. Med.,* 61, 294, 1968.
22. Huntley, C. C. and Stevenson, R. E., Maternal phenylketonuria. Course of two pregnancies, *Obstet. Gynecol.,* 34, 694, 1969.
23. Johnson, C. C., Phenylketonuria and the obstetrician, *Obstet. Gynecol.,* 39, 942, 1972.
24. Perry, T. L., Hardwick, D. F., Lowry, R. B., and Hansen, S., Hyperprolinaemia in two successive generations of a North American Indian family, *Ann. Hum. Genet.,* 31, 401, 1968.
25. Mollica, F., Pavone, L., and Antener, I., Pure familial hyperprolinemia: isolated inborn error of amino acid metabolism without other anomalies in a Sicilian family, *Pediatrics,* 48, 225, 1971.
26. Bruckman, C., Berry, H. K., and Dasenbrock, R. J., Histidinemia in two successive generations, *Am. J. Dis. Child.,* 119, 221, 1970.
27. Cunningham, G. C., Two years of PKU testing in California. The role of the laboratory, *Calif. Med.,* 110, 11, 1969.

INDEX

A

B